Beekeeping – From Science to Practice

Russell H. Vreeland · Diana Sammataro

Editors

Beekeeping – From Science to Practice

 Springer

Editors
Russell H. Vreeland
Bickering Bees Farm
Belle Haven, VA
USA

Diana Sammataro
Tucson, AZ
USA

ISBN 978-3-319-86904-9   ISBN 978-3-319-60637-8   (eBook)
DOI 10.1007/978-3-319-60637-8

Printed on acid-free paper

This Springer imprint is published by Springer Nature
The registered company is Springer International Publishing AG
The registered company address is: Gewerbestrasse 11, 6330 Cham, Switzerland

This book is dedicated to hard working pollinators everywhere and to all of those working to protect and help them!

Preface

Many people enjoy hobbies simply to relax or, in some cases, perhaps use them as ways to make extra money to support their avocation. Most of them will readily admit that part of the joy of their outside activity is learning about it, becoming an expert in the field, and delving deep into the wells of knowledge that surround every worthwhile pursuit. Beekeepers are constantly peppered with information about their hives, bees, parasites, diseases, and all types of management techniques from numerous experts at all levels. They are also inundated by questions from caring and well-meaning non-beekeepers. Part of the problem from all sides of this issue then is having quality, up-to-date information to use in both beekeeping and public events. Naturally, there is plenty of anecdotal information about everything and lots of opinions. It is sometimes said that if you ask five beekeepers a question you will certainly get ten answers and seven opinions. We have faced these same issues both as beekeepers and as scientists.

In assembling *Beekeeping — From Science to Practice*, we attempted to find some of the best most active scientists currently studying honeybees worldwide. We asked them to provide information not only on their personal research areas but also clear precise generalized information that would benefit beekeepers. While this current book is by no means comprehensive, it does present diverse chapters discussing problems with Queen rearing, developing diseases, pests, pesticides, epidemiology, and the value of products (i.e., propolis) and even aspects of newly developing strains of European Honeybees. Bees are declining, and the reasons are complex, but with the information in this work, beekeepers will better understand the intricacies of this highly social insect and what we know right now. In turn, we hope the information will help beekeepers keep their bees healthy and thriving in the face of the huge challenges from today's world. We and the scientists who agreed to contribute to this book hope that a better understanding will also allow us to all to comprehend and solve the challenges that all pollinators will face in the coming decades. Also in assembling this material, we tried to concentrate on providing some in-depth coverage of issues that are often only mentioned in general honeybee manuals, and in a couple of cases, we included some straight scientific information that while not directly usable in an apiary may well lead to new methods (and bees) in the not too distant future.

We have had an exceptional time assembling these authors, reading their chapters, and learning many new things ourselves. We hope that you the reader benefit from these insights as much as we did. We also hope this information will be useful to you and will help your beekeeping operations.

Belle Haven, USA   Russell H. Vreeland
Tucson, USA   Diana Sammataro

Contents

About the Editors

Dr. Russell Vreeland is a chauvinist about two things: microorganisms and pollinators of all types. As a microbiologist, he knows that it is really microbes that run the Earth while the rest of us often just mess it up. He initially trained as a marine microbiologist in the early 1970s and ended up studying the remains of the ancient oceans (large underground salt deposits) as a geological microbiologist. During this part of life, he focused entirely on the marvelous microbes that survive and thrive in saline waters anywhere from 2x to 10x the concentration of seawater.

Dr. Vreeland's first experience with hardworking pollinators came in the late 1990s when he obtained his first 12 tubes of solitary pollinator bees. Within three years, he was supplying these little darlings to neighbors, friends, and farmers near his home in Pennsylvania as well as to his small wildlife refuge on the Eastern Shore of Virginia. At one point, he and his wife estimate they maintained well over 150,000 solitary pollinators. The solitary pollinators eventually gave way to a sweet tooth, and he started keeping honeybees in 2004. Dr. Vreeland met Diana Sammataro at breakfast when she applied for a professorship at West Chester University a friendship that has now lasted for almost 18 years. In 2007, Drs. Vreeland and Sammataro began a joint collaborative study to examine the microbial population and changes in bee breads in healthy hives. After both retired from active basic research, they decided to collaborate on this book.

Dr. Vreeland and his wife Susan currently own "Bickering Bees Farm" located in Craddockville, VA. He is active in the local Beekeepers Guild of the Eastern Shore, frequently gives talks about all types of bees (honey and solitary pollinators) to numerous local civic organizations and schools and maintains enough hives to sell honey and make mead. Russell has also continued to pursue his other professional love in the form of a small business called Eastern Shore Microbes using his amazing salt-loving microbes to treat and eliminate highly saline wastewaters from industries all over the world. When he has spare time, he and his dog "Beesley" go fishing or work with the local US Coast Guard Auxiliary.

Diana Sammataro Co Editor of Beekeeping—From Science to Practice is also the co-author of the Beekeeper's Handbook. She began keeping bees in 1972 in Litchfield, Connecticut, setting up a package colony in her maternal grandfather's old bee hive equipment. From then on, she decided that her B.S. in Landscape

Architecture (University of Michigan, Ann Arbor), would not be a career, but that honey bees would. After a year of independent studies on floral pollination (Michigan State University Bee Lab, East Lansing), she earned an M.S. in Urban Forestry (University of Michigan, Ann Arbor). In 1978 she joined the Peace Corps and taught beekeeping in the Philippines for 3 years. On returning, she worked at the USDA Bee Lab in Madison, Wisconsin, under Dr. Eric Erickson, studying the effects of plant breeding and flower attraction of bees in sunflower lines. When the lab closed, she eventually went to work at the A.I. Root Company as Bee Supply Sales Manager in Medina, Ohio. In 1991, she was accepted at the Rothenbuhler Honey Bee Lab at The Ohio State University, Columbus, to study for a Ph.D. under Drs. Brian Smith and Glen Needham. In 1995, she worked as a post-doctoral assistant at the Ohio State University Agricultural Research Center in Wooster, Ohio, with Dr. James Tew and in 1998 at the Penn State University Bee lab. Early in 2002, she was invited to join the USDA-ARS Carl Hayden Honey Bee Research Center in Tucson, Arizona. She worked there as a Research Entomologist until 2014, looking at bee nutrition problems, how they influence Varroa mites, and current pollination problems

What We Learned as Editors

Russell H. Vreeland and Diana Sammataro

Abstract

In assembling a book like this one, the editors often learn as much, or more, from the exercise than most people believe. Each of the chapters has taught us important lessons some of which include things like the immense value of propolis for bees. So we need to encourage production rather than grumble at the glue. The need for really good record keeping comes through in several chapters as does the value of these records when combined with similar data from other areas. While every modern beekeeper is likely anti-pesticides the wide ranging negative effects on bees come through loud and clear in several chapters. But where most focus is on simply the death of the insects, our contributors show that even "harmless" pesticides/fungicides are attacking everything from rearing quality queens to the nutritive value of the beebread produced in the hives and how "sub-lethal" doses alter larval development and survival. Speaking of larva neither of us knew that there are really two forms of American Foulbrood that attack our hives. One kills quickly and is often found, while the second type takes longer and may be unrecognized until it is too late. We learned how far Small Hive Beetle larva will travel to reach soil and how truly devastating these predators really are once they get into a hive. Then there is the material about the well-known *Varroa* mites, how they have become stronger and how they have become vectors that have increased the strength of specific viruses through a process in which viruses combined in the mite, or the cues used by the mite to infect the larva. On the positive side though we found out about the growing number of Honeybee strains that are fighting back against the mites (some we didn't know about), we also learned about the new developments in honeybee

R.H. Vreeland (✉)
Bickering Bees Farm, 10460 Teackle Rd., Belle Haven, VA 23306, USA
e-mail: rvreeland@wcupa.edu

D. Sammataro
DianaBrand Honey Bee Research Services LLC, Tucson, AZ, USA

1

cell cultures that will allow a closer better study of the viruses. Finally, we discovered aspects of venom allergies their frequency and risk factors for all us beekeepers. So there is a lot to learn here in these pages, for us and for you.

There are several reasons behind a decision to assemble a book like this one. We recognized that as scientists, we spend a lot of time conducting experiments and developing important data that helps our understanding of much of Earth's natural phenomena and the biota that makes it so interesting. The first problem, however, is that scientists are specialists and tend to focus on specific topics. Second, in order to communicate effectively with each other, scientists often use technical language and terminology. Third, scientific knowledge is generally published in a wide variety of specialized technical journals that have a very limited distribution or availability. So if beekeepers want to get the most up-to-date honeybee science, they must hunt down the journals that have the specialized papers and become conversant in highly technical language, then find out which scientists are doing what type of research. On the other hand, the scientists can spend some time assembling their information into a straightforward discussion directed at the beekeepers who need the information. Truthfully, there is a fourth reason; as scientists and beekeepers ourselves, we realized that the previous three reasons pertained as much to us as they did to the non-scientist beekeepers. We wanted to know what our many colleagues understood, what we in our own specialties did not have time to follow and we wanted to see if we could use all of this valuable information in our own hives. Hence, this book was assembled to provide a source of reference that any beekeeper may use to find the best management practices for any beekeeping operation.

One of the most important lessons that can be gained from examining each (and all) of these chapters is a recognition of the difficulty scientists face in trying to understand these complex and beautiful creatures. Time and again throughout this text, the readers will see that strong and clear results obtained in one set of studies are not quite so strong and clear when the experiments (or observations) are repeated in a subsequent season. This is not necessarily due to "poor science" as some might quickly assume. Rather, every attempt, no matter how carefully constructed, is complicated by the presence of a different group of bees, different equipment, weather, or other circumstances. Biology is hard to control.

As we assembled this book, we read and edited each of these chapters and we learned a tremendous amount that we intend to use in our respective operations. As a way to help direct you the reader, we decided to discuss what we found to be the most important, most interesting (or in some cases most technical) material coming out of each of these chapters. We have also attempted to provide an overall synthesis with advice on what we see as the best ways to apply this information in beekeeping operations of nearly any size.

We wanted to identify what we see as the important information from each chapter and to stress that what we present here is not all of the information provided in each chapter. This chapter represents a basic summary of the things we enjoyed and what we learned the most in each of the chapters.

1 Propolis

For some unknown reason, bees today produce less propolis and do not coat the equipment like they do in a tree cavity. But propolis helps keep out pests and diseases. Borba et al. (2017) describe the results of studies on propolis usage by Africanized bees. They discuss the fact that the colonies with lots of propolis produce stronger, more abundant brood, have workers with longer life spans, store more honey and pollen, and have better overall hygiene. Most of us are aware that hives use propolis to seal up cracks and reduce the flow of cold air into the hives in winter. In addition, Borba et al. (2017) discuss the fact that in healthy hives, propolis serves even more functions that include acting as an antimicrobial agent and an overall disinfectant for the entire superorganism. Apparently, this function is only a property of fresh propolis and does not last over winter. That means the hive must replace propolis every year.

One of the more interesting aspects they discuss is the fact that the bees do not seem to put much propolis in finished wood used in modern hives. In studies cited by Borba et al. (2017), the rough interiors of natural hives (those in trees and other structures) were covered with large amounts of propolis. The images in the propolis chapter illustrate that in order to get the bees to collect enough propolis for the studies, the researchers had to cover the insides of the boxes with propolis traps.

Overall, Borba et al. (2017) show that hives that produce more propolis appear to have more and healthier bees and are better hives throughout the seasons.

How can beekeepers make use of this information?

1. Maybe we should attempt to get our bees to produce more propolis. We can do that by being sure they have access to the types of trees that produce these resins.
2. Perhaps we should (as is actually recommended by Borba et al. (2017)) stop doing so much sanding and smoothing of the inner hive body surfaces (or all boxes) in order to stimulate propolis collection.

2 Pesticides

Lundgren (2017) holds modern-day pesticides with the same level of esteem as Rachel Carson did over 50 years ago. That is to say outright banning their production and use is too good for them. In this chapter, he makes an eloquent case for all of the problems they cause to everyone. In reality, whether or not one uses these chemicals, everyone is exposed to them at some level. According to a sustainability Web site (www.sustainabletable.org), there are currently 350,000 toxic pesticides approved for use in the USA. To keep this in perspective, the United Nations currently estimates that there are about 15,000 nuclear warheads in the entire world; the US Army has fired 250,000 bullets for every insurgent killed, and since 1827, the US Food and Drug Administration has approved a total of 1423 drugs. The point here being that we have turned our world upside down having far more ways to kill living things than we do to keep them alive and healthy. This is not a sustainable situation.

In his chapter, Dr. Lundgren discusses the cost/benefit of using so many pesticides on our food crops and our environment. He also discusses some of the value judgements society imparts, such as working harder or paying more to save a species once they are threatened than to simply keep from threatening them in the first place. He also illustrates the reality that we beekeepers are as much to blame for our heavy use of Coumaphos and Fluvalinate when *Varroa* mites first appeared in European honeybees.

Throughout this discussion, Dr. Lundgren points out the many hidden problems with beeswax, some of which practicing beekeepers do not generally consider. Some of these problems include the fact that the numerous materials added to pesticide formulations often react synergistically with one another and become toxic in their own right. In some instances, manufacturers add several "inactive" ingredients to a single formula, thus creating entirely new product lines that are still toxic. As most of us know, these pesticides are soluble in the wax and can last for years. Dr. Lundgren (2017) points this out and, in addition, questions the fate of these toxins when we create, sell and burn beeswax candles, creams or make food wraps impregnated with beeswax. Are we in fact spreading these pesticides into our own foods, on our skin, and the air in our homes? Does the burning or heat in wax melters make the chemicals (and its mixtures) more toxic, break it down to something less harmful, or, even worse, release it into the atmosphere as a gas? None of this is known, and few are considering it.

Dr. Lundgren's chapter also provides a good overview of the different modes of action of the pesticides in general, and shares some examples of each group and how each might attack our hives. He discusses the different ways in which our bees are exposed both inside and outside of the hive and at what life stages they are most impacted.

How can beekeepers make use of this information?

1. We must at every opportunity advocate for a saner approach of Integrated Pest Management in Agriculture. Many states are establishing guidelines to protect all types of pollinators. It is up to us, as beekeepers, to advocate for language that requires use of biological pest management practices (and many exist) over the selection of synthetic chemical pesticides.
2. We must advocate for better toxicity screening (one that takes all aspects of hive biology into account) before approval of new chemicals.
3. As for our own operations, we need to be more diligent at removing older wax (which has the highest residues) from our hives. There is no information on whether or not this old comb is toxic, so decisions on using it for other things have to be made on the spot. Getting wax tested at an USDA laboratory is also encouraged.

3 Queen Quality

DeGrandi-Hoffman and Chen (2017) discuss the scientific focus on producing high-quality queens for the honeybee industry. This is clearly a critical aspect for everyone for many reasons. Without a quality strong queen in the colony, we face sequential supercedures which ultimately lead to the demise of the hive. We also face problems of weakening genetics, as virgin queens can mate with related drones. In reality, much of the problem described here is closely related to information in other chapters about pesticides (see chapters by Yoder et al. (2017) and Lundgren (2017)). From an editorial point of view, this was not done on purpose. These topics (other than that by Lundgren 2017) were not solicited for the purpose of discussing pesticides. In every case, the authors were free to present their own material and all three ended up discussing pesticide issues; this points to the problems these chemicals are still causing more than 50 years after the appearance of Rachel Carson's "*Silent Spring.*"

DeGrandi-Hoffman and Chen (2017) also addressed at least one aspect of external pesticides not discussed by others; the effects of two compounds when these are combined within a hive are combined. In this case, the information is about the insecticide Chlorpyrifos (Lorsban by Dow Chemical) combined with the "harmless to bees" fungicide "Pristine®." For those who have never heard of this, it is a combined fungicide featuring two different chemicals. It is made by BASF and is approved for use on a wide variety of fruits and vegetables, including grapes, strawberries, virtually all stone fruits (peaches, nectarines, apricots, and cherries), pome fruit (apples and pears), tree nuts (especially almonds, but includes pecans, chestnuts), and carrots and other bulb vegetables (onions, etc.). So this is a very common chemical in our world. Chlorpyrifos is listed as an insecticide, miticide, and acaricide, so it hits just about anything without a backbone. They also found that when these two combine in the hive, multiple things occur; first of all, fewer queen cells survive to the hatching stages. Those queen cells that do not hatch have dead larva that resemble those killed by Black queen Cell Virus. In all cases, queen larvae that were exposed to only a single pesticide hatched at a significantly higher rate than did those exposed to the pesticide and the fungicide.

Overall, DeGrandi-Hoffman and Chen show us why so many of our supercedures seem to fail, why our queen suppliers are having trouble supporting us, and lastly perhaps why our queens are simply not as good as they once were.

How can beekeepers make use of this material?

1. In our present era, pesticides and fungicides are literally everywhere. They are used in homes and yards throughout suburbia, and in urban and rural environments. Folks use multiple chemicals because they do not want to pull weeds, or they try to rid themselves of roaches. Even flea and tick collars are now coming complete with neonicitinoids and last 8 months. So about the best we can do is speak up and speak out. The public is concerned about bees and their losses but they do not realize that these materials really do not work. Beekeepers need to lend their voices, at every opportunity.

2. As a group, we must stop using synthetic miticides, antibiotics, and chemicals in our hives.
3. When we suspect hives are killed by chemical poisons, we must at least attempt to notify State Apiarists.

4 Bee Bread and Fungicides

Everyone now recognizes the negative impact of pesticides on honeybees and other native pollinators. Now we are learning more about the supposedly safe fungicides. As discussed, DeGrandi-Hoffman and Chen (2017) determined the synergistic effects of a fungicide combined with a pesticide on queen rearing. Yoder et al. (2017) bring the fungicides into a different and otherwise ignored part of the beehive, the pollen that is stored to become bee bread. Before launching into a brief discussion of these effects, it might be advantageous to understand why this is important. Many of our basic beekeeping books talk about bees eating pollen (or feeding it to the developing larva); if that were the entire story, it might not be too bad. But as with everything else in beekeeping, there is a back story that Yoder et al. (2017) are just beginning to address. The back story is that stored pollen does not just sit, it ferments and becomes "bee bread." If you have the chance to taste some of it, it tastes like sourdough. Like all fermentations, the process both preserves and enhances the nutritional quality of the fermented material. A significant part of that process (generally the initial fermentations) is carried out by fungi. That means that fungicides will have an impact on the fermentation process of converting pollen into bee bread.

Yoder et al. (2017) have now shown that on the one hand, the fungicides attack the beneficial fungi that help the hive ferment pollen to form highly nutritious bee bread. At the same time, these chemicals alone and in combination (see DeGrandi-Hoffman and Chen 2017) create a situation that makes hives more susceptible to viruses and Nosema disease, while also reducing their ability to raise viable quality queens. So two things come out of this: First, beekeepers now need to be even more vigilant about hive locations relative to spraying as even "safe" chemicals are not so safe. Second, the common denominator for all three things— bee bread quality, pathogen susceptibility, and queen quality—is overall nutrition.

How can beekeepers make use of this material?

1. Stop using synthetic chemicals on or around your hives. Just stop.
2. Advocate for more regulation on all agricultural sprays and work to educate those around you. Get your clubs to speak out.
3. Advocate for better testing of new sprays so that safe is really safe and not something just hidden from view.
4. Get used to the smell of your healthiest hives; you can smell the sweet gases of good fermentations and know that things are going well.

5. Notice if the pollen in your hive is being used or ignored. If the latter, remove it —there is something wrong.

5 Cell Cultures

Goblirsch (2017) presents a discussion of some of the newest and most powerful techniques being developed to study CCD on the molecular level, that is, investigating the DNA, proteins, and other aspects in the cells of honeybees. In the past few decades, molecular biology has increased our understanding of all creatures, even humans. It gives us the tools, for example, to detect and isolate differences in proteins in healthy versus diseased bees. In working with complex organisms such as honeybees having sustainable, reproducing cell lines is literally the first step in helping us select the genes that can strengthen our bees and enable them to fight off things like mite-transmitted viruses discussed elsewhere in this book. Alternatively, it could enable us to breed bees that are resistant to foulbrood at all stages not simply as adults.

This is a relatively new field of honeybee science because, up until the parasitic mites were found in Europe, and later in the USA, there seemed to be no need to attempt the difficult and time-consuming experiments needed to generate such cell cultures for these beneficial insects. However, with the disruption of both the mites and the newer *Nosema ceranae* disease present in bees, the development of new techniques to study these problems give us a powerful tool to help understand and perhaps treat such threats. Insect cells are more of a challenge to culture (then say human cells) and were used to help understand pest species such as flies, leafhoppers, or moths, not to study beneficial insects. New techniques allow researchers to now study the effects disease, nutrition, and pests have on bees in a more controlled way (e.g., no weather, diet, pest stress fluctuations) and to determine how bees would respond at cellular and molecular levels.

A good example is the new Nosema disease now of grave concern to all beekeepers. This global fungal infection can now be studied without infecting hives that could spread the disease. Tightly controlled laboratory-based cellular studies can now be used to determine how the spores of *N. ceranae* are transferred, how they invade host cells, and to even detect different strains of Nosema.

New cell lines are also being used to study the multiple viruses now infecting bees, including bumblebees and the interactions between virus and hosts. Again, these can be performed in the laboratory, without the danger of an escape or having the infectious agents spread to innocent hives. In time, these studies will help in developing therapies for these viruses. Other uses of this technology include cell lines from nervous tissues that could study the effects of toxins (e.g., neonicotinoids) and potentially from midgut tissues. All these approaches will give researchers and beekeepers tools for helping our important pollinating insects.

How can beekeepers use this material?

1. In truth, normal beekeeper cannot make direct use of cell lines such as these. The editors decided to include this material as an example of some of the great breakthroughs that may come out of future research efforts.
2. What beekeepers can do is voice their continued support for research funding for these studies, and they are definitely not inexpensive but as they develop, their power and application will grow exponentially.
3. What can be done, however, is that local clubs with a sufficient savings can (and perhaps should) seek out research like this near you and provide even small amounts of support. Even a thousand dollars will purchase many of the chemicals and laboratory expendables needed to perform these studies.

6 Viruses

Chejanovsky and Slabezki (2017) have provided a thorough examination of the many viruses that are now adding to the problems faced by beekeepers (and by bees) all around the world. At the start of the infestation, colonies of *A. mellifera* were able to sustain high levels of *Varroa* infestations. Mites, up to 10,000 per colony, were not lethal. Now, mite levels above 3000 per colony may be enough to cause colony collapse (Boecking and Genersch 2008).

It is now apparent that least two viruses carried by *Varroa* can (and have) combined *with one another so that each virus has* become more virulent in infecting honeybees. That means it is more critical than ever to maintain at least some level of control over these mites. While we should not try to eliminate them with materials that just make the mites more resistant, beekeepers must not let these ectoparasites get too numerous in their hives at any time of the year. More frequent "soft" treatments that keep the mite numbers down may be better than a single "hard" blast. A better alternative to any type of treatment regimen may be to keep naturally hygienic strains of bees as much as possible (a topic to be discussed below).

Many beekeepers want to allow nature to act, with the goal of developing greater resistance in the bees. This is certainly a worthy effort; however, the development of many *Varroa* transmitted viruses makes this a dangerous situation not just for your own hives but also for all other beekeepers in the area. Certainly if a beekeeper lived on an island and was the only beekeeper there, this would be fine. However, today, this could promote the rise of more virulent viruses (and possibly even new viral combinations), making life even harder for honeybees.

Chejanovsky and Slabezki (2017) also make one additional point that many may not consider. Bees, when stressed by any number of factors, simply become more susceptible to viral attack. In fact, one of the things that happens is that viruses that are present in the hive, and are asymptomatic, are being held in check by the bee's immune systems and basic cleanliness. However, if the hive becomes stressed, that situation changes and the viral disease appears.

One of the things that becomes obvious from reading this chapter will be the reality that these viruses are highly contagious and can be transmitted pretty easily. How can beekeepers make use of this material?

1. Unfortunately, it is impossible to look into a hive and find the viruses. All we can do is to maintain as clean a hive as possible, especially when it comes to *Varroa* mites. Too many of us assume that hives do not have many mites or we decide on our own to try to raise our own resistant bees. That may not be a good idea anymore.
2. Obviously, it is important that we try to limit the spread of these viruses. One way to accomplish that is to clean heavily used hive tools. The same can be said for taking material from a dying hive and putting it into a healthy hive. If the sick hive has viruses, you just transferred them. You can go from healthy to sick, not the reverse.
3. If a hive dies and you do not know why, clean or replace the wooden ware (or isolate it for months or scorch with fire) before reuse. Do not just reuse the stuff, if you need it clean it.

7 Epidemiology

The chapter by VanEnglesdorp and Steinhauer (2017) is possibly one of the more scientifically focused chapters presented here. However, the material is extremely important for beekeepers, especially in this time of colony losses. Epidemiological information is highly statistical since it attempts to deal with large numbers over even larger geographical areas. This is necessary to understand and to evaluate disease levels in colonies.

This chapter outlines the study of epidemiology, or disease levels in a population in a clear way that can be useful to anyone who keeps bees, or even significant numbers of other animals. The chapter describes epidemiology as a discipline, what measurements are taken, and what risk factors are taken into account. It also provides clear examples of how epidemiologists conduct experiments and attempt to track disease outbreaks. One of the compounding difficulties in following such parameters with honeybee colonies (unlike say illnesses in horses) is that beekeepers do not check hives daily (sometimes not even weekly) so the principle measures are not always freshly obtained. Further, unlike the sudden death of a larger system, beekeepers cannot perform autopsies to confirm cause of death.

While not every beekeeper will go into such details, it is important that everyone keeps good records of their colonies and their health, medications, and disease incidents. By understanding how disease (and mites) impacts our colonies and what we can do to offset or control them, we can better identify risk factors in our colonies before these factors overwhelm the colony. This is especially important when selecting colonies as potential queen breeders to incorporate into a program for rearing honeybee strains that are resistant to diseases and mites. In addition, keeping track of what our bees are exposed to, and their mite loads, will also keep beekeepers alert to the introduction of new pests/diseases, such as the new hive beetle and the yet to be discovered (in the USA at least) *Tropilaelaps* mites. By monitoring our bees, we will have a better chance of preventing diseases from spreading and mites from overrunning our colonies. By keeping track of colony losses, we as beekeepers can also help reinforce the need for more research funds to help our bees survive. If any other agricultural industry (beef, poultry) sustained an average of 21% (or higher) loss, there would be immense research monies instantly available to determine causes and remedies; we need this for our pollinators.

How can beekeepers use this material?

1. Carefully examine your colonies, get to know what a healthy colony is like. If you do not understand what normal looks, smells and sounds like, you cannot detect what is out of place or wrong.
2. Keep careful records. They do not necessarily need to be detailed but if they tell you how a hive was doing at time 1 you can quickly recognize a difference at time.
3. Participate in surveys both national and local. If your club does not conduct a survey urge that one be started. Find out how many in your area lost bees, how much honey was produced, or generally how local bees are faring. In this case, local beekeepers are the literal first line of defense in controlling problems and protecting bees.

8 Small Hive Beetles (SHB)

Dr. Pirk (2017) provides a thorough discussion of the biology and the problems being caused by this pest. In their native habitat of southern Africa, *Athenia tumida* (the small hive beetle) is a relatively minor problem. But like all invading species once it was carried out of that area and found a new target (this time the European honeybee), it became a major economic threat. Now hive beetles can be found throughout the world invading managed and feral colonies of honeybees, and we now realize that they are a worse problem. As discussed by Dr. Pirk, these beetles and especially their larval stages can truly devastate even healthy hives.

A huge part of the problem of dealing with SHB is their life cycle much of which occurs inside the hives. Even though the bees chase them, beetles hide, congregate, and then mate in small cracks and areas where bees cannot reach. One female beetle can lay over 300 eggs in a day. Larva that hatch, then eat honey, pollen and even the young honeybee brood after which the beetle larva are attracted to each other and congregate into a few cells. Many beekeepers only see them at this point, but the damage is already done. The beetle larva attack and feed on the brood and stores for anywhere from 3 to 21 days depending on how much they eat. Consequently, when a hive is doing well and is fully stocked you get more generations of beetles. Then, the beetle larva attempt to leave the hive, enter the soil, and become adults. In reality, it is only at this stage that the beetles are easily susceptible to control (except when you can squish them with a hive tool).

Dr. Pirk describes several aspects of control of these pests both from the perspective of the bees (a strong hive can attack the adults but also identify the young larva and drop them outside where they die; a weakened hive cannot) and from that of the beekeeper. Beetles, like honeybees, are insects so insecticides cannot be used, but Dr. Pirk provides a good review of the numerous traps available and even a hint of some possible future control mechanisms. One of the disturbing stories he relates shows that beetle larva can travel over 30 m (33 yards) to reach soil so ideas about setting hives on plywood, plastic, or even small areas of cement just do not work.

How can beekeepers use this material?

1. Be vigilant and if you see a hive beetle do not assume that is the only one—take some action.
2. Nematodes in the soil do work and should be spread in an area around 30 cm from the hive.
3. 85% of the larva will pupate in the top 6 inches of the soil with the rest in the top foot.
4. Reducing the hive entrances will help to reduce the access by the beetles. Use traps at the top and the bottom of the hive to catch the adults.

9 American Foulbrood

Foulbrood is probably the single problem most feared by beekeepers. As discussed by Genersch (2017), it was first recognized by Aristotle over 2000 years ago. It has been plaguing beekeepers even before that. Foulbrood not only devastates one's personal hives but also requires testing all hives in the area as well. Further, there really is no cure save destroying all of the infected hives and doing some really heavy duty cleansing on all woodenware. Controlling foulbrood is really the issue that started modern beekeeping inspection programs.

Dr. Genersch provides an excellent review of the history of foulbrood through its past scientific relevance and how it defied identification for many years. This was due in large part to the fact that the causative agent *Paenibacillus larvae* grows so poorly on virtually all microbiological media. This organism only infects the larva in a hive within the first 36 h after they hatch from the egg. As Genersch (2017) points out, a single infected larva can contain billions of pathogenic microbes and especially spores once the larva dies. These spores are picked up by the resistant adults when they attempt to clean dead larva. The spores are then transmitted to other hives or to more susceptible larva by the nurse bees.

Dr. Genersch (2017) points out several aspects of this process. First, the pathogen actually has two life stages: first as a pathogen during which it attacks the larva and the second as a saprophyte during which it degrades the dead larva forming that sticky ropy mass and scales in the cells. These are of course the clinical symptoms (in addition to the smell that all beekeepers hope to never experience) that beekeepers and inspectors look for in hives. As pointed out (Genersch 2017), there is no cure other than destroying the infected bees and hives. Things such as antibiotics only keep the infection in check. This generally works only as long as the *Paenibacillus* is susceptible to the antibiotic; once it becomes resistant (and there are resistant forms), the infection will run rampant again.

Possibly, the novel and most important thing discussed by Dr. Genersch (2017) is that there are actually two forms of AFB. These were identified using genetic testing systems and are known as ERIC 1 and ERIC 2. These two forms both kill the hive but they do so at different rates and generally only one will be detected by the known clinical symptoms. Space does not permit us to describe these, and Dr. Genersch does a much better job of it than either of us are able.

How can beekeepers make use this material?

1. All beekeepers should at least become used to the sweet smell of a healthy hive. If that changes, dig deeper and quickly.
2. In the more virulent AFB, most of the larvae die before metamorphosis and cell capping begin. In the second slower acting form, 90% of the deaths will occur post-capping but once that level is reached the hive is finished.
3. The disease and virulence differences between these two AFB forms is counterintuitive. The more rapid form is more devastating for individual larva (it kills them faster) but because of that the hive's social immunity and cleanliness will prolong the demise of the colony.
4. The second AFB form is less devastating for individual larva BUT because the larvae are capped prior to death, this form is more destructive for the colony as a unit.
5. Bottom line—nothing beats a vigilant beekeeper, with a good understanding of his/her hives. You may not be able to save one but you may likely save the many.

10 Pheromones

Nearly every beekeeper will understand the importance of pheromones in the life of honeybees. In short, it is one of their primary ways of communication with their sisters in a dark hive. Honeybees use these simple chemicals to make sure the queen is healthy and has plenty of sperm stored, and they use pheromones to identify hive mates, to identify proper and related eggs (or even remove extra drone eggs), to know the life stage of a hive mate, and even to guide one another home during foraging.

Critically, honeybees also use pheromones to know when larva need feeding or needs to be capped for metamorphosis. Part of the problem is that *Varroa* mites are exposed to these pheromones and make use of them to know when to enter larval cells. Dr. Y'ves LeConte is one of the world's pheromone experts who has spent a lifetime identifying and studying this interaction. Dr. Le Conte's chapter (LeConte 2017) outlines some of the important work he has done on the identification of pheromones in honeybee colonies. In an effort to understand what chemical signals the *Varroa* mites can detect, his laboratory started to examine different pheromones produced by bee larvae. Using hexane solvent, they extracted compounds from the larvae. They then used rather simple but scientifically elegant experiments and identified which of the numerous extracted compounds attracted the mites.

They also found that these compounds, not unexpectedly, had an effect on other bees, such as signaling the nurse bees to cap larval cells. During these and other experiments, Dr. LeConte and his students identified numerous other compounds that had many other effects within the hive environment. These pheromone compounds caused the bees to increase royal jelly production and to delay the age at which workers leave the hive to go foraging. A good working knowledge of how these compounds alter and control nearly all bee behaviors is a critical aspect for beekeepers trying to reduce stress on hives or to aid their bees in survival. Dr. LeConte does add a word of caution, however, noting that while some of these chemicals can be fed to bees to manipulate their behavior, making them too concentrated or applying them at the wrong times could have significant adverse effects. Since we do not understand truly how these compounds work, their entire range of effects, or even the full suite of pheromones produced by these insects, a lot more research needs to be done.

How can beekeepers use this material?

1. Dr. LeConte names many of the compounds he has identified. Beekeepers can look these up, learn their smells, and simply use the nose to help understand their hives.
2. Recognizing that smoke disrupts these important materials, we can use this tool judiciously and teach new beekeepers proper smoking techniques and how much they should use.
3. While it might be tantalizing to try to apply synthetic pheromone-like compounds to hives to help control behaviors, this would not be a recommended practice.

11 Varroa-Resistant Bees

Throughout this book, *Varroa* mites have probably been mentioned nearly as often as honeybees, probably because they have become so intimately associated with those far more beneficial insects. As has often been noted, beekeepers worldwide have been doing many things to control these pests, some (as in North America) resorting to chemical controls (both hard and soft), others to a mix of home remedies. But the problem is, this often yields stronger mites and weaker bees. This is particularly obvious in the statistics that when *Varroa* first appeared a hive could survive with as many as 10,000 mites per hive while many present-day hives succumb with only 3000 mites per hive.

As pointed out in other chapters, mites have developed an additional and most troublesome aspect. That is, they are now known to be vectors of viruses that can be devastating for our honeybees. One of the most telling aspects in this regard has been the increased virulence of several viruses (especially deformed wing virus) that were once relatively minor problems in hives but are now (after passage through *Varroa*) more problematic than ever.

Now, as discussed by LeConte and Mondet (2017), the bees may have a fighting chance. That is, evolution (with a little help from science) is producing numerous strains of *Varroa*-resistant European honeybees. These bees include those that attack and remove mite infected larva (called *Varroa* Specific Hygienic [VSH] bees); strains in which the mites experience poor reproduction (Suppressed Mite Reproduction [SMR] bees); bees that attack and bite off mite legs and other body parts (so called Biter bees) and even to strains that collect propolis containing large amounts of mite inhibiting compounds (called caffeic acids).

LeConte and Mondet (2017) also provide some directions for obtaining these resistant bees as part of an overall control program that avoids the problems of sprays and chemical treatments.

How can beekeepers use this material?

1. Search for and purchase bees that show a natural resistance to mite attacks. They will allow you to decrease sprays, etc.
2. If you have these strains be willing to spread them around.
3. Recognize that this kind of development is the absolute best way to approach the mite problems. These honeybee strains are interfering with the mites' natural life cycle, which means that this ultimately leads to a peaceful coexistence. We will never get rid of mites but at least we can help the bees manage the problem to our (bees and beekeepers) mutual benefit.

12   Venom Allergies

Allergic reactions to honeybee venom are quite literally an occupational hazard. We may keep bees for 50 years and get stung consistently over that time without incident but as Drs. Ricketti and Lockey (2017) point out severe allergic reactions can happen to anyone at any time after the first exposure. In their chapter, Ricketti and Lockey provide an excellent description of the differences between a large local reaction (LLR) and systemic allergic reaction (SAR). Plus, they provide pictures for those who may not know. They point out that while only 1–4% of the total population experiences a systemic reaction to bee stings, nearly 31% of beekeepers experience them. Further, and perhaps more sobering, is that 30–60% of all beekeepers have the venom specific IgE antibodies that are associated with these severe reactions.

Ricketti and Lockey (2017) also discuss the reality that while beekeepers receive the most stings during their first year of beekeeping their risk of responding with a severe reaction increases the longer they are beekeepers. Another aspect that bears notice, is that the risk of having an SAR also increases in beekeepers that demonstrate an allergic response to other common allergens and who tend to develop allergic rhinitis, asthma, or even skin reactions (allergic dermatitis). They also point out that while significant numbers of beekeepers demonstrate an allergic reaction only about 18% of those carry a lifesaving epinephrine injectable while handling bees.

This chapter provides a great, easy-to-understand discussion of the biological differences between honeybee and wasp venom, and it also gives a nice overview of the different stages in a reaction and the different levels of reactions. This is something all beekeepers need to understand for yourself and for the many people you meet who say they are allergic to bee stings because their hand swells when stung on the finger. That is a local reaction. It would be systemic if the hand swells when they are stung on the foot.

How can beekeepers use this material?

1. By all means if you keep bees, consider asking your physician (or immunologist) for a prescription for self-administered epinephrine. Then, carry it with you in the bee yards; it might be the only thing that can save your life.
2. Be aware of stings and do not think it is a sign of bravery to avoid them. Beekeeping is fun, that is why we do it, but the reactions are not. In fact, one of us (RHV) gained firsthand knowledge during his second year keeping bees but still has them along with an injectable.
3. Protective gear is exactly that, it is protective and if you have multiple allergies use it all, wear the gloves and jacket. It is more than a good idea it just might save your life.

References

Boecking O, Genersch E (2008) Varoosis—the on-going crisis in beekeeping. J Consum Prot Food Saf 3:221–228

Borba RS, Wilson MB, Spivak M (2017) Hidden benefits of honeybee propolis in hives. In: Vreeland R, Sammataro D (eds) Beekeeping—science to practice. Springer: Switzerland

Chejanovsky N, Slabezki Y (2017) Honey bee viruses- Pathogenesis, mechanistic insights and possible management projections. In: Vreeland R, Sammataro D (eds) Beekeeping—science to practice. Springer: Switzerland

DeGrandi-Hoffman G. and Y. Chen. (2017) Sub-lethal effects of pesticides on queen rearing success. In: Vreeland R, Sammataro D (eds) Beekeeping—science to practice. Springer: Switzerland

Genersch E (2017) Foulbrood diseases of honey bees—from Science to Practice. In: Vreeland R, Sammataro D (eds) Beekeeping—science to practice. Springer: Switzerland

Goblirsch M (2017) Using Honey Bee Cell Lines to Improve Honey Bee Health. In: Vreeland R, Sammataro D (eds) Beekeeping—science to practice. Springer: Switzerland

Le Conte Y (2017) Discovering a honey bee brood pheromone. In: Vreeland R, Sammataro D (eds) Beekeeping—science to practice. Springer: Switzerland

Le Conte Y, Mondet F (2017) Natural selection of honey bees against *Varroa destructor*. In: Vreeland R, Sammataro D (eds) Beekeeping—science to practice. Springer: Switzerland

Lundgren JG (2017) Predicting both obvious and obscure effects of pesticides on bees. In: Vreeland R, Sammataro D (eds) Beekeeping—science to practice. Springer: Switzerland

Pirk CWW (2017) Small hive beetles (*Aethina tumida* Murray) (Coleoptera: Nitidulidae). In: Vreeland R, Sammataro D (eds) Beekeeping—science to practice. Springer: Switzerland

Ricketti PA, Lockey RF (2017) Honeybee venom allergy in beekeepers. In: Vreeland R, Sammataro D (eds) Beekeeping—science to practice. Springer: Switzerland

Steinhauer N (2017) Using epidemiological methods to Improve honey bee colony health. In: Vreeland R, Sammataro D (eds) Beekeeping—science to practice. Springer: Switzerland

Yoder JA, Nelson BW, Jajack AJ, Sammataro D (2017) Fungi and the effects of fungicides on the honey bee colony. In: Vreeland R, Sammataro D (eds) Beekeeping—science to practice. Springer: Switzerland

Author Biography

Dr. Russell Vreeland is a chauvinist about two things: microorganisms and pollinators of all types. As a microbiologist, he knows that it is really microbes that run the Earth while the rest of us often just mess it up. He initially trained as a marine microbiologist in the early 1970s and ended up studying the remains of the ancient oceans (large underground salt deposits) as a geological microbiologist. During this part of life, he focused entirely on the marvelous microbes that survive and thrive in saline waters anywhere from 2× to 10× the concentration of seawater.

Dr. Vreeland's first experience with hardworking pollinators came in the late 1990s when he obtained his first 12 tubes of solitary pollinator bees. Within 3 years, he was supplying these little darlings to neighbors, friends, and farmers near his home in Pennsylvania as well as to his small wildlife refuge on the Eastern Shore of Virginia. At one point, he and his wife estimate they maintained well over 150,000 solitary pollinators. The solitary pollinators eventually gave way to a sweet tooth, and he started keeping honeybees in 2004. Dr. Vreeland met Diana Sammataro at breakfast when she applied for a professorship at West Chester University, a friendship that has now lasted for almost 18 years. In 2007, Drs. Vreeland and Sammataro began a joint collaborative study to examine the microbial population and changes in bee breads in healthy hives. After both retired from active basic research, they decided to collaborate on this book.

Dr. Vreeland and his wife Susan currently own "Bickering Bees Farm" located in Craddockville, VA. He is active in the local Beekeepers Guild of the Eastern Shore, frequently gives talks about all types of bees (honey and solitary pollinators) to numerous local civic organizations and schools, and maintains enough hives to sell honey and make mead. Russell has also continued to pursue his other professional love in the form of a small business called Eastern Shore Microbes using his amazing salt-loving Microbes to treat and eliminate highly saline wastewaters from industries all over the world. When he has spare time, he and his dog "Beesley" go fishing or work with the local US Coast Guard Auxiliary.

Hidden Benefits of Honeybee Propolis in Hives

Renata S. Borba, Michael B. Wilson and Marla Spivak

Abstract

Honey bees (*Apis mellifera* L.), like many social insects, have collective behavioral defenses called "social immunity" to help defend and protect the colony against pathogens and parasites. One example of social immunity is the collection of plant resins by honey bees and the placement of the resins on the interior walls of the nest cavity, where it is called a propolis envelope. Propolis is known to have many antimicrobial proprieties against bacteria, fungi, and viruses and has been harvested from bee hives for use in human medicine since antiquity. However, the benefit of propolis to honey bees has not been studied until recently. This chapter focuses on how bees collect and use the antimicrobial properties of plant resins within the hive as a form of social immunity and defense against infectious bacterial and fungal pathogens. The studies presented here demonstrate the significance of the propolis envelope as a crucial component of the nest architecture in honey bee colonies. The collection and deposition of resins into the nest architecture impact individual immunity, colony health, and support honey bees' antimicrobial defenses. These studies emphasize the importance of resin to bees and show that plants are not only a source of food, but can also be "pharmacies."

R.S. Borba (✉) · M.B. Wilson · M. Spivak
University of Minnesota, 530A Hodson Hall; 1980 Folwell Avenue,
St. Paul, MN 55108, USA
e-mail: rsborba@umn.edu

© Springer International Publishing AG 2017
R.H. Vreeland and D. Sammataro (eds.), *Beekeeping – From Science to Practice*,
DOI 10.1007/978-3-319-60637-8_2

17

1   Benefits of Propolis to Colony Health

It is common knowledge that honey bees forage for pollen, nectar, and water. What is not well appreciated is that honey bees also forage for plant resins, but not for nutritional reasons. Resin is a sticky exudate secreted by plants to protect young leaf buds or the entire plant from disease, UV light, and herbivore attack (Langenheim 2003). Resins are composed primarily of antimicrobial compounds (e.g., terpenes and flavonoids) that play a major role in the defense and survival of the plant (Langenheim 2003). Many animals, including bees, collect these antimicrobial resins for their own health benefits. In bees, the presence of resin in the nest plays a major role in the immune defense of individual bees, improving colony health and fitness (Simone et al. 2009; Simone-Finstrom and Spivak 2012; Borba et al. 2015; Borba, 2015).

Honey bees collect resin mainly from buds and leaves of various tree species, but they also collect resins from droplets appearing on the trunks or limbs of trees (Alfonsus 1933), and from a few tropical flowers (Kumazawa et al. 2003; Armbruster 1984). Bees can extract resin by fragmenting leaves with their mandibles (mouthparts) or collecting it directly from the plant surface (Meyer 1956; Teixeira et al. 2005). Bees collect resins to varying degrees; some honey bee species and races use resins extensively, such as the African-derived subspecies *Apis mellifera scutellata* and the European-derived subspecies *A. mellifera caucasica*. At least one species of honey bees, *Apis cerana,* does not collect resin (Butler 1949; Page and Fondrk 1995). In colonies that do collect resin, the number of resin foragers depends on the needs of the colony (as discussed later in this chapter), but generally they comprise less than 1% of the total forager work force. Resin collection is a very difficult and time-consuming task to perform. After chewing pieces of resin from the plant, bees must transfer the sticky secretion from their mandibles to their hind legs before returning to the hive. Because of the sticky characteristics of resin, once back in the hive, resin foragers need the assistance of other bees to remove the resin load from their legs, which may take up to 30 min (Fig. 1; Nakamura and Seeley 2006). The bees will then carry the resin in their mandibles to the site in the hive where the resin will be deposited. Once deposited in the nest, the resin, sometimes mixed with beeswax, becomes what beekeepers know as propolis.

Honey bees naturally nest in tree cavities where they coat the entire inner surface of the nest cavity surrounding the combs with a propolis envelope (Seeley and Morse 1976). Seeley and Morse (1976) suggested that the propolis envelope had various functions, including serving as an impermeable barrier to tree sap and environmental moisture, a solid surface for comb attachment, a physical barrier to outside invaders by sealing the holes and cracks of the nest cavity, and finally, an antimicrobial layer against natural occurring fungi and bacteria in the tree cavity. When nesting in a hollow tree cavity, honey bees prepare the new nest site by removing the soft, rotten wood from the nest walls and depositing propolis in the cracks to make it solid and smooth (Seeley and Morse 1976). Beekeepers,

Fig. 1 Worker bee removing resin load from the hind leg of a resin forager. Upon return to the hive, resin foragers need the assistance of other bees to remove the resin load from their legs, which may take up to 30 min (*Photo credit* Christine Kurtz)

particularly in the USA, have selected against colonies that collect large amounts of propolis (Fearnley 2001) because its stickiness makes opening and managing colonies in standard beekeeping equipment difficult. Importantly, honey bees do not construct a propolis envelope in standard beekeeping equipment because the inner walls of the wooden boxes are already solid and smooth, which apparently does not stimulate propolis deposition. Instead, bees deposit propolis in dispersed cracks and crevices in manmade hive bodies and not as a continuous envelope as they do within a tree cavity (reviewed in Simone-Finstrom and Spivak 2010).

Honey bees are very resilient insects; they have thrived in this world for 6–8 million years (Engel 1999), relying only on their own natural defense mechanisms to survive. Although propolis has been used as a traditional and natural human medicine since biblical times (Simone-Finstrom and Spivak 2010), the benefits of propolis for honey bee health were not appreciated until the last decade. Studies have demonstrated that the presence of a propolis envelope enshrouding the nest area is a fundamental component of honey bee colony health (Simone et al. 2009; Simone-Finstrom and Spivak 2012; Borba et al. 2015; Borba and Spivak, *in review*). The propolis envelope functions as an antimicrobial, or "disinfectant" layer around the nest, and thus as an external layer of the colony immune defense. This chapter will summarize current research since the previous review (Simone-Finstrom and Spivak 2010), emphasizing research conducted by R. Borba: (1) the seasonal benefits of a propolis envelope to colony health and individual honey bee immunity; (2) the role the propolis envelope plays in bees' natural defense against brood diseases; and (3) how honey bees select and use plant resins as a form of self-medication.

1.1   Seasonal Benefits of Propolis to Bee Immunity and Colony Health Under Natural Field Conditions

A honey bee colony can be considered a superorganism, a group of related individuals living together in a nest with the ability to perform collective foraging, thermoregulatory and defensive behaviors. When collective behavioral mechanisms are used to defend the colony against parasites and pathogens, they are called mechanisms of social immunity (Cremer et al. 2007). Examples of social immunity in honey bees include hygienic behavior (the ability of adult bees to detect and quickly remove diseased and mite-infested brood from the nest, thus limiting pathogen and parasite transmission; reviewed in Evans and Spivak 2010), grooming (removal of the parasitic *Varroa* mite from a nestmate's body; Boecking and Spivak 1999), and foraging for resins to construct a propolis envelope inside the nest (Simone et al. 2009; Simone-Finstrom and Spivak 2012).

The benefits of the propolis envelope to honey bee health were first investigated by coating the inside of managed hives with a propolis extract (solution of 13% propolis in 70% ethanol) using a paintbrush and exposing bees to this propolis-enriched environment for 7 days (Simone et al. 2009). After one week, 7-d-old bees had lower immune system activation and lower bacterial loads in and on their bodies compared to same-age bees in hives without the propolis-extract coating (Simone et al. 2009). This short-term study indicated that bees in hives with the propolis-extract envelope did not have to expend as much energy turning on (activating) their immune system to fight off microbes in the nest. When the immune system of bees, or any animal, is activated, it comes with a physiological cost (e.g., reduced survival; Moret and Schmid-Hempel 2000). In fact, for insects the immune system may be the most costly physiological system to maintain (Evans and Pettis 2005; Schmid-Hempel 2005). When the immune system does not need to be highly activated, as when there is a propolis envelope in the nest cavity, bees may be able to allocate that saved energy to perform vital tasks (e.g., foraging and rearing brood) or store protein in their bodies.

Recent research from Brazil showed that Africanized bee colonies that deposited high amount of propolis had greater brood viability, longer worker lifespan, higher honey production, more rapid hygienic behavior, and larger pollen stores compared to colonies that deposited low amounts of propolis (Nicodemo et al. 2013, 2014). Even though most European-derived stocks of bees in the USA do not deposit much propolis, it is possible that they also receive the same *long-term* benefits from the antimicrobial compounds in propolis. Therefore, a research experiment was conducted to investigate the long-term benefits of a propolis envelope, but this time testing propolis naturally collected and deposited by the bees inside the nest. Colonies were encouraged to build a natural propolis envelope by cutting and stapling commercially available propolis traps to the four inner walls of each hive box in 12 colonies (propolis envelope treatment group; Fig. 2). The bees readily filled the 24×3 mm (height $\times$ length) gaps in the traps with resin they collected from the field. No propolis traps were provided to another set of 12 colonies, and the bees deposited propolis in the cracks and crevices within the box only where

Fig. 2 Propolis envelope treatment bee box. **a** Propolis traps stapled to inside walls of a hive to encourage bees to construct a propolis envelope. **b** View of the propolis envelope when traps were removed at the end of the experiment. In each colony, the bees deposited propolis within most of the gaps of each propolis trap (*brown lines on the box* are the deposited propolis). In a tree cavity, the propolis envelope is contiguous, but bees do not tend to deposit propolis on planed wooden walls in beekeeping equipment, unless lumber is left unfinished and very rough

they could (control group). This experiment was conducted on a first set of colonies from April 2012 to May 2013 and was repeated on a new set of colonies from April 2013 to May 2014. Each year the colonies were started from package bees on unused equipment and combs.

During the active foraging season (from July to September) and the following May of both years, the following measures were taken on *colony* health: (1) adult bee population size, (2) total amount of worker brood, and (3) levels of *Varroa* mites and *Nosema* spp. Adult bee populations were estimated in each colony by counting the number of frames covered with bees in each box (following Nasr et al. 1990). Worker brood was quantified by placing a 2.6 cm^2 grid over each frame and counting the number of squares filled with sealed or unsealed brood (following Nasr et al. 1990). *Varroa* levels were measured by collecting samples of 300 adult bees from the brood area and dislodging the mites from the bees in the laboratory (following Lee et al. 2010; Spivak and Reuter 2001a). *Nosema* levels were measured by counting *Nosema* spp. spores in 100 bees using a hemocytometer (Cantwell 1970).

The presence of a propolis envelope in the colony did not appear to have an effect on adult bee population size as colonies with and without a propolis envelope had similar adult bee populations over both replicated years of the experiment. There was a potential effect of the propolis envelope on the brood area; as in the first replicate of the experiment, colonies with a propolis envelope had significantly larger brood areas in May 2013 compared to the colonies without the propolis

Worker brood population size

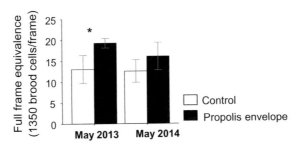

Fig. 3 A bar graph with standard errors is used to represent the worker brood population size in May 2013 and May 2014. The average (±standard error) number of full frame equivalents (1350 worker brood cells) is shown on the y-axis and the months are indicated on the x-axis. Significant differences between controls (*white*) and propolis envelope (*black*) treatment colonies are indicated with * (*P* < 0.05)

envelope. A similar trend was observed in May 2014, but the difference in brood areas was not statistically significant (Fig. 3). It was predicted that the propolis envelope might lower the levels of pathogens and parasites (Nosema spp. and parasitic mites) in the colonies. However, in both replicates of the experiment, these levels were very low and did not differ between the colonies with a propolis envelope and control colonies with no propolis envelope, likely because all colonies began as "packages" and pathogen and parasites levels do not usually rise to high levels in new colonies the first year in Minnesota. Thus, all colonies in the experiment were apparently healthy.

The effects of propolis on *individual* bee health were measured in 7-d old bees by quantifying: (1) the levels of three common viruses (DWV—Deformed Wing Virus, IAPV—Israeli Acute Paralysis Virus, and BQCV—Black Queen Cell Virus), (2) the expression of specific immune genes, and (3) the level of a blood storage protein called Vitellogenin (Vg). All three individual bee health measurements were quantified using a common (but somewhat expensive) laboratory technique called real-time, quantitative PCR (polymerase chain reaction; see explanation of this technique below).

There were no significant differences in levels of all three viruses (DWV, IAPV, and BQCV) between colonies with a propolis envelope and those without. The lack of high levels of viruses, in addition to low levels of *Varroa* and *Nosema*, support the hypothesis that colonies from both treatments were apparently healthy. Further studies will be necessary to explore the effect of propolis on viral levels, as well as other bee pathogens and parasites, when colonies are highly infected.

When infected with a pathogen, bees and humans can initiate an immune response via cellular or humoral immune pathways. The cellular immune response includes the engulfing and encapsulation of pathogens by blood cells, while the humoral immune response includes the production of small antimicrobial proteins that attack and kill pathogens. The starting point of humoral immune system activation is called gene transcription, when a particular sequence of DNA (a gene) is transcribed to make messenger RNA (mRNA), which "message" is then translated to make a specific protein. It is possible to measure how much mRNA is being produced of a particular gene using real-time, quantitative PCR. After extracting mRNA from an individual bee, or a group of bees, one needs to backtrack the natural order of gene transcription (from DNA to mRNA) and reverse-transcribe mRNA (single-stranded sequence of a gene) to DNA (double-stranded sequence of the same gene), which can then be used in PCR reactions. PCR amplifies a specific double-stranded sequence of genes into billions of copies, which are then quantified. A specific sequence of genes is targeted using "primers," which are small segments of nucleotides that are complementary to a piece of the gene that is to be amplified. For the experiments described here, primers were designed to target honey bee immune-related genes, specific viruses (DWV, IAPV, and BQCV), and Vitellogenin.

To measure immune system activation, it is best to collect bees of the same age, preferably young nurse bees, as the immune systems of older foragers become highly variable in expression levels. To collect young bees, newly emerged bees were paint-marked with a dot of enamel paint on the thorax just after they crawled out of their cells. Six days later, when bees were 7-d old, 25 paint-marked bees were collected from each of the experimental colonies. Bees were collected in the summer, fall and the following spring in both years to measure the immune system activity.

Measures of immune gene activation revealed that bees within the colonies with naturally constructed propolis envelopes had significantly lower immune gene expression, or much "quieter" immune systems, over the summer and fall months compared to bees in control colonies. In fact, bees in colonies with a propolis envelope had less variable (more uniform) immune gene expression over the active foraging season (Borba et al. 2015). A decrease in energetic costs associated with the maintenance of an efficient immune system may help bees to allocate their energy to perform vital tasks (e.g., foraging and rearing brood) and to maintain higher storage protein levels (e.g., vitellogenin) required for overwintering success.

Immune gene expression data is often shown in scientific journals using bar graphs that are somewhat difficult to interpret. Figure 4a shows results for only one of the six immune genes measured from 7-d-old bees in September 2012 and September 2013, and the figure legend explains how to interpret the graph. Another way to show the data is through a visual representation where the low-to-high levels of gene expression are represented as colors, called a heat map. Figure 4b shows a heat map representation of the expression of all six immune genes for September 2012 and September 2013.

Surprisingly, by the following spring of both years, before the bees were actively collecting resin again, there were no significant differences in gene expression levels for most immune genes between bees from the two treatment groups (Fig. 5). This finding suggested that the bees' immune systems were not benefitting from the propolis envelope in early spring. To solve this conundrum, it became important to explore the possibility that the propolis deposited by bees in the previous summer and fall had lost some of its antimicrobial activity over the winter. Using a test

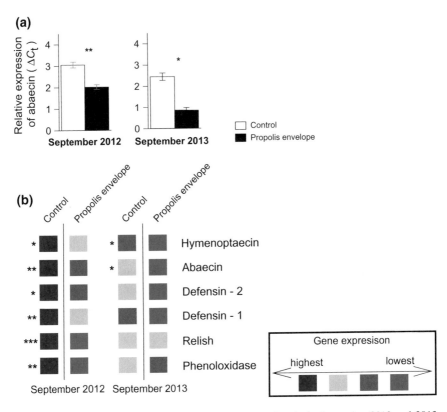

Fig. 4 a Relative expression levels of the immune gene abaecin in September 2012 and 2013. The expression levels are shown relative to the expression of reference genes Actin and RPS-5 ($\Delta Ct = (\bar{x}(\text{reference genes})\, Ct - \text{target gene }Ct)$) Reference genes are not involved in immunity but are produced in relatively equal amounts by bees over their lifetime to regulate other physiological functions. When the immune gene expression is high, the value on the vertical y-axis is a higher number. The height of each bar indicates the average (mean) value of the data for each immune gene, and the lines extending upward and downward from each bar represent the variation, standard error, around the mean. The *white bars* represent the control colonies and the *black bars* represent the colonies with a propolis envelope. **b** In this modified heatmap, the different colors represent low and high levels of gene expression. The color *blue* represents the lowest values of gene expression, followed by *green, yellow* and *red*, the highest level of gene expression. Values were statistically compared between treatment groups (propolis envelope and control colonies) for each gene separately (and thus colors can be compared only for each gene separately). Significant differences in gene expression between treatments (when results are considered statistically different) are indicated by * with increasing number of *'s indicating a higher probability of being different: * = $P < 0.05$; ** = $P < 0.01$; *** = $P < 0.001$). A P value greater than 0.05 indicates that there is no difference between the two treatment groups, while a P value lower than 0.05 means that the two treatments are significantly different

described later in this chapter (see Sect. 1.3), it was found that, indeed, the propolis within the nest in late April had lost much of its antimicrobial activity from the previous fall. The loss of biological activity of the propolis from October to April is probably due to the lack of new resins being brought in over the winter. Honey bees

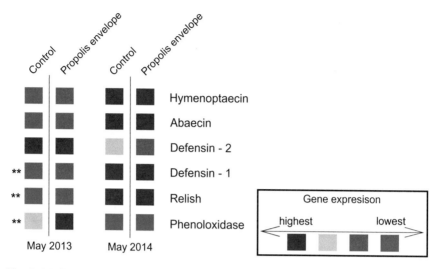

Fig. 5 Modified heatmap graph of the relative expression of immune genes in May 2013 and 2014. Significant differences in gene expression between treatments (when results are considered statistically different) are indicated by "**" ($P < 0.01$). A P value greater than 0.05 indicates that there is no difference between the two treatment groups, while a P value lower than 0.05 means that the two treatments are significantly different

in Minnesota do not forage for resin (or for any resources) during the cold temperature season, from October to April, but start collecting resin again later in May, when environmental temperatures for tree growth are favorable. With the deposition of new propolis in the nest in the spring and throughout the growing season, the benefits of the propolis envelope to the bees would return.

Even though the bees' immune systems did not benefit from the propolis envelope in early spring, measures of blood storage protein, vitellogenin (Vg), were significantly higher in spring of both years (May 2013 and May 2014) in bees from colonies with a propolis envelope compared to bees from control colonies (Fig. 6). Vg is an important protein in bees' hemolymph, and when present in high concentration is an indicator of well-nourished bees (Amdam et al. 2003, 2004; Engels et al. 1990). More recently, it has been found that Vg also contributes important priming function to the immune system of bees (Salmela et al. 2015). This high Vg level in bees from propolis envelope colonies in the spring of both years suggest that these bees had more protein storage compared to bees in control colonies and, therefore, is a possible explanation for why they were able to rear more brood compared to the control colonies (Bitondi and Simoes 1996; Mattila and Otis 2006).

The results from this experiment provided information on the long-term benefits of propolis to honey bee health, adding important new information to that reported by Simone et al. (2009).

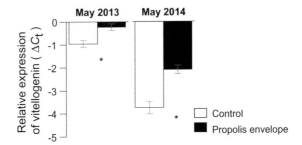

Fig. 6 Average (±standard error) of relative expression levels ($\Delta Ct = (\bar{x}(\text{reference genes})\, Ct - \text{target gene } Ct)$) of vitellogenin. As in Fig. 2a, a higher value on the y-axis means higher expression (e.g., –2 is higher than –4). Vitellogenin levels were higher in bees from propolis envelope colonies (*black*) compared to control colonies (*white*) in the spring of both years. Significant differences between controls and propolis envelope treatment colonies are indicated with * ($P < 0.05$)

In sum, colonies that are allowed to construct a natural propolis envelope on the inside of the hive boxes benefitted in ways that improve bee health and possibly colony strength and survivorship. The propolis envelope creates an antimicrobial layer around the bees that, remarkably, serves as an environmentally derived component of the bee's immune defense. Propolis could help bees' immune system either by reducing the microbe load in the nest cavity, as suggested by Simone et al. (2009), or by having a direct and beneficial effect on bees' immune system (Borba et al. 2015).

A human analogy. To fully understand the function and benefits of the propolis envelope to bees, it is helpful to draw an analogy between a honey bee nest and human homes. Mold and fungi are often found in our houses, especially during spring and summer when humidity is higher. The presence of these microorganisms in the air may not always cause a health problem, but some people's immune systems are easily affected by these microorganisms, and the inhalation of molds and fungi can lead to immune activation in more sensitive people. The propolis envelope to bees would be the same as coating the walls of our homes with an antimicrobial material. In that case, the antimicrobial material would effectively decrease the levels of microorganisms growing on the walls of the house and indirectly prevent our immune system from activating an immune response (Simone et al. 2009). Even in the absence of high levels of molds and fungi, the presence of the antimicrobial propolis could directly decrease the need to express immune genes (Borba et al. 2015). Lower immune system activation (the immune system at an efficient "idle") does not imply immune suppression (the immune system turned off; see Sect. 1.2). The lower immune system activation is beneficial because mounting a strong immune response comes with a cost. The immune system needs to use energy to fight off pathogens and when it is always activated, the individual, whether bee or human, is left with less bodily resources and greater immune stress, which may affect overall health and ability to fight off secondary or subsequent infections.

1.2 The Role of the Propolis Envelope for Bees' Natural Defense Against Brood Diseases

In addition to the everyday (constitutive) benefits of the propolis envelope to the bees' immune system (as described in Sect. 1.1), the antimicrobial properties of propolis supports honey bee natural defenses against pathogens. A recent study found that honey bee colonies coated with a propolis extract (experimentally applied envelope) and challenged with *Ascosphaera apis*, a brood fungal pathogen that causes chalkbrood, had less chalkbrood infected brood compared to challenged colonies with no propolis-extract coating (Simone-Finstrom and Spivak 2012). Colonies with a propolis-extract coating had an average of 14.7 ± 7.5 chalkbrood infected larvae per colony, while challenged colonies with no propolis coating had an average of 108.2 ± 49.0 chalkbrood infected larvae per colony. The mode of action by which the propolis decreases clinical signs of chalkbrood in honey bee colonies is not yet understood, but these initial findings were intriguing and led to another study to test the effect of a natural propolis envelope on a different bee disease: American foulbrood.

American foulbrood (AFB) disease is caused by the bacterial pathogen, *Paenibacillus larvae*. American foulbrood is highly infectious to honey bees and can rapidly spread among colonies via drifting (when a forager enters a colony that is not their own) and robbing of contaminated nectar. Young honey bee larvae (1–2 d old) are highly susceptible to this pathogen, while old larvae and adults are considered resistant. A potential reason for this susceptibility is thought to be because young larvae have "less developed" immune defenses compared to older brood and adults (young larvae have lower bee "blood" cell counts and cellular defense mechanisms; Chan et al. 2009; Wilson-Rich et al. 2008).

Previous studies have demonstrated four different mechanisms of colony resistance to AFB: (1) removal of *P. larvae* spores from contaminated honey by the filtering action of the proventricular valve between the bee's crop and ventriculus (stomach; Sturtevant and Revell 1953); (2) detection and rapid removal of AFB-infected brood by adult bees before the pathogen becomes infectious (hygienic behavior; Spivak and Reuter 2001b); (3) genetic ability of larvae to resist AFB infection (Evans 2004; Rothenbuhler and Thompson 1956), and (4) ability of nurse bees to secrete antimicrobial compounds into larval food, which can protect the larvae somewhat from *P. larvae* infection (Rose and Briggs 1969; Thompson and Rothenbuhler 1957). Additionally, numerous laboratory studies have demonstrated that propolis has antimicrobial properties that inhibit the growth of *P. larvae* (Bastos et al. 2008; Bilikova et al. 2013; Wilson et al. 2013, 2015). Therefore, the next experiment explored whether the antimicrobial activity of a natural propolis envelope could support bees' natural mechanism of defense against AFB.

Three questions were posed: (1) After challenging colonies with the bacterium that causes AFB, would the level of immune genes be higher in nurse-age bees in colonies with a propolis envelope compared to nurse-age bees in colonies without the envelope? (2) Would the antimicrobial activity of larval food supplied by nurse bees to young larvae be higher in challenged colonies with a propolis envelope? And (3) would there be less AFB-infected brood in colonies with a propolis envelope?

In the summer of 2013, ten colonies were stimulated to construct a propolis envelope by stapling propolis traps to the inner walls of standard beekeeping boxes (as explained in Sect. 1.1). Five of the ten colonies were experimentally challenged with *P. larvae* by spraying a sugar solution with a known concentration of *P. larvae* spores on each comb within the colony (propolis + *P. larvae* treatment). The other five colonies with a propolis envelope were left unchallenged (propolis + no *P. larvae* treatment). Another set of ten colonies was not provided with a propolis envelope and the bees deposited propolis in the cracks and crevices within the box where they could. Similarly, five of the ten colonies without a propolis envelope were challenged with *P. larvae* (no propolis + *P. larvae* treatment) and the other five were left unchallenged (no propolis + no *P. larvae* treatment).

Samples of 7-d old bees were collected to test the expression levels of immune genes (as explained in Sect. 1.1), once before and once after challenged colonies showed clinical signs of AFB (August 9 and September 12, respectively). Samples of larval food were collected to test its antimicrobial activity. Larval food from 1- to 2-d old larvae was collected on the same day as the 7-d-old bees were collected (asymptomatic period August 9, and symptomatic period September 12). Prior to larval food collection, an empty frame was introduced into the colony and was marked when eggs were present. Three days after the frames were marked, when 1-2-d old larvae were present, the frames were removed and larval food was collected following Schmitzová et al. (1998). In a temperature-controlled room, each young larva was removed from the cell using a sterile grafting tool, and the larval food from each cell was individually homogenized in 30 µl of phosphate buffer by repeated pipetting and then transferred to individual tubes.

The number of larvae with clinical signs of AFB (sunken wax capping and uncapped cells containing discolored, ropy brood) on each frame of each colony was quantified approximately every 15 days after the appearance of the first clinical sign (August 30, September 16 and October 1).

The antimicrobial activity of larval food was measured in liquid culture. Most bacteria, such as *P. larvae*, can be grown under controlled laboratory conditions, in tubes containing a liquid with the required nutrients for bacterial growth (called broth). Bacterial growth in liquid culture is characterized by the increased turbidity of the culture, and the optical density (OD) of the liquid culture can be measured using a spectrophotometer. This machine produces a light of a preselected wavelength in one end of the chamber that houses the sample, and records the intensity of light detected at the other end of the chamber after it passes through the sample. Samples with greater concentrations of bacteria have a greater optical density and will absorb more light, reducing the intensity of light that reaches the detector. Therefore, the intensity of the light detected decreases as the sample concentration of bacteria, and optical density, increases. The antimicrobial activity assay consisted of allowing a known concentration of a *P. larvae* culture (pre-grown in brain/heart infusion broth for 48 h prior to the assay) to grow in the presence of larval food for

6 h at 37 °C and subsequently evaluating the bacterial growth by measuring the optical density (OD at time 0 h subtracted from time 6 h). Bacterial growth was compared in cultures with added larval food relative to cultures without added larval food (controls).

Immune gene expression analysis of nurse-age bees collected after the appearance of AFB clinical signs showed that bees from challenged colonies with a propolis envelope had a stronger immune response compared to bees in challenged colonies without a propolis envelope, as indicated by significantly higher gene expression levels of two antimicrobial peptides (hymenoptaecin and apidaecin). It is well known that honey bees increase the expression of most antimicrobial peptides, including hymenoptaecin and apidaecin, to fight a *P. larvae* infection (Chan et al. 2009; Evans 2004). However, *P. larvae* spores do not germinate in adult bees and therefore do not cause any harm to nurse-age bees. Thus, the inducible physiological response of nurse-age bees to AFB infection may not be to protect adult bees against this pathogenic infection but to protect young larvae that are fed by them, which are highly susceptible to this disease. These gene expression results indicate that nurse bees from propolis envelope colonies have the ability to synthesize higher levels of antimicrobial peptides and potentially decrease colony-level AFB infection more rapidly and efficiently compared to bees in challenged colonies without a propolis envelope. Importantly, these findings also demonstrate that bees in colonies with a propolis envelope are able to mount a strong immune response after they are challenged. Thus, the lower immune system activation ("quieter" immune system) of bees in apparently healthy colonies with a propolis envelope (see Sect. 1.1) is *not* due to immune suppression (i.e., the inability to mount an immune response), because after challenge these bees are able to quickly activate their immune responses.

Nurse bees perform the behavioral task of feeding the brood by regurgitating larval food into the cells and therefore are in constant direct contact with the susceptible larval stage to AFB. We found that when challenged colonies had a propolis envelope, the bioactivity of the larval food was significantly higher compared to the larval food in unchallenged colonies without a propolis envelope. The higher antimicrobial activity of larval food in challenged colonies with a propolis envelope suggests that antimicrobial compounds from the propolis envelope may contribute directly to the bioactivity of larval food against bee pathogens. Although the propolis envelope may not come into direct contact with larval food, volatile compounds present in propolis can diffuse through the hive and may contribute to the complex way in which bees fight infections. Another hypothesis is that nurse bees in challenged colonies with a propolis envelope that produce more antimicrobial peptides, incorporate these antimicrobial peptides (Bilikova et al. 2001) into larval food fed to 1–2 d old larvae to increase young larvae immune defense mechanism to fight *P. larvae* infection. Either way, these results confirm the existence of a natural defense mechanism in honey bees against AFB by feeding larvae food with a higher antimicrobial activity (Rose and Briggs 1969; Thompson and Rothenbuhler 1957). Importantly, both mechanisms of defense against AFB (higher immune gene expression and larval food bioactivity) were only observed when challenged colonies had a propolis envelope.

Clinical signs of AFB can be identified by the presence of sunken wax cappings and uncapped cells containing discolored, ropy brood. As a measure of the level of AFB infection, the number of cells containing signs of AFB was counted in each comb (Spivak and Reuter 2001b). A severity score ranging from 0 – 3 was given for each comb (both sides combined) that contained larvae: 0 = 0 cells containing signs of AFB; 1 = 1-5 cells; 2 = 6-25 cells; and 3 = ≥ 26 cells per comb (Spivak and Reuter 2001b). An overall AFB severity score for each colony by each month (i.e., August, September and October) was obtained by calculating the median (± interquartile range) of the individual comb scores (Table 1). These results indicate that the presence of a propolis envelope inside a colony reduced the number of larvae with clinical signs of AFB over time, but did not eliminate the disease completely. The reduced level of AFB clinical signs in early October in colonies with a propolis envelope compared to colonies without a propolis envelope is likely a result of a combination of the effects of propolis on both the collective and individual behavioral responses (larval food bioactivity and individual bee immune response), as well as the incorporation of antimicrobial peptides by nurse-age bees into larval food. The study by Simone-Finstrom and Spivak (2012) on the effect of a propolis-rich environment on the infection level of chalkbrood disease reported that colonies with a propolis-extract coating inside the nest had a level of infection 86% lower (14.7 ± 7.5 cells compared to 108.2 ± 49.0) than observed in colonies without the propolis extract. Similarly, the findings presented here show that colonies with a propolis envelope had 52% fewer cells infected with AFB in October compared to colonies without a propolis envelope (Table 1).

To summarize, the presence of a propolis envelope increased the individual and collective immune responses of bees, possibly by supporting the increased production of antimicrobial peptides in individual nurse bees, and increasing bioactivity of larval food fed collectively by nurse bees. As a result, AFB clinical signs in early October in colonies with a propolis envelope were reduced compared to colonies without a propolis envelope. The propolis envelope served as an external antimicrobial layer around the colony, protecting the brood from *P. larvae* infection and supporting bees' ability to induce a strong and effective immune response with the result of a lower infection load after two months following the challenge.

Table 1 AFB infection level data was measured by counting the number of cells containing signs of AFB in each comb.

Treatment	Number of colonies	AFB clinical sign (median ± interquartile range)		
		August	September	October
No propolis envelope + *P. larvae*	5	0.875 ± 1.187	1.5 ± 1	2.429 ± 0.863
Propolis envelope + *P. larvae*	5	0.625 ± 1.125	0.125 ± 1.875	1.167 ± 0.733
Statistical significance		$P > 0.05$	$P > 0.05$	$P = 0.036$

The median number of total AFB-infected cells were compared between treatments. A P value lower than 0.05 indicates a statistically significant difference between the two treatment groups

1.3   Do Bees Self-Medicate?

Self-medication is defined as the "defense against pathogens and parasites by one species using substances produced by another species" (Clayton and Wolfe 1993). If bees can truly self-medicate, an individual (or colony) should perform a behavior, such as resin collection, at higher rates when parasitized and at lower rates when healthy. Simone-Finstrom and Spivak (2012) found that honey bee colonies increase resin foraging after exposure to chalkbrood, revealing that bees medicate the colony with resin in response to this particular fungal infection. To extend the knowledge of how honey bees exploit resin to fight pathogen infection, a recent study by R. Borba investigated whether bees also self-medicate in response to a bacterial infection, American foulbrood (AFB).

This study was repeated over 3 years from 2012 to 2014, using new sets of colonies each year. Colonies equalized in population size and food resources were used, and resin foraging activity was monitored when the colony was healthy and after experimentally challenging them with either *P. larvae*, the causative agent of AFB (in 2012, 2013 and 2014), or *Ascosphaera apis,* the causative agent of chalkbrood (CB; in 2014 only). The number of resin foragers was assessed before pathogen challenge by closing the colony entrance once or twice a day (weather depending) for 15 min between 1100 and 1600 h for 12 observation periods (spread over two weeks) and recording the number of foragers returning with a resin load on the hind legs. After 12 observations, one group of colonies was challenged with a *P. larvae* spore solution (using the same methods described in Sect. 1.2), and the second group of colonies served as controls (unchallenged colonies). In 2014, a third group of colonies was provided with a pollen patty containing *A. apis* spores. Resin foragers were again counted over another set of 12 observations periods spanning two weeks. The change in resin foraging between the pre-challenge and post-challenge periods was calculated for each colony by subtracting the total number of foragers before challenge from the total number of foragers after challenge, and this difference was compared among treatment groups (control, AFB- and CB-challenged colonies).

The results showed that colonies challenged with *P. larvae* have a slight numerical increase in resin foraging in 2012, 2013, and 2014 compared to unchallenged colonies (Borba, 2015). In 2014, bees from CB-challenged colonies had a substantial and statistically significant increase in resin foraging, as they did in the study by Simone-Finstrom and Spivak (2012).

Do bees self-medicate with specific plant sources of resin? When people are sick, they can go to the pharmacy and self-medicate by buying an over-the-counter drug that treats the infection they are experiencing (e.g., bacterial or fungal infection). Honey bees self-medicate in a similar way by collecting antimicrobial resins ("drugs") from plants ("pharmacy"), but it is not known if bees choose specific resins that are most able to treat the infection the colony might have.

Chemical composition of resins varies qualitatively and quantitatively within and among plants (Witham 1983). Wilson et al. (2013) conducted a study on the bioactivity of resins from 14 tree species against *P. larvae* growth. The resins were collected from trees on the St. Paul campus of the University of Minnesota, and the findings revealed a significant difference among botanical sources of resins to inhibit the growth of this bacterium. Likewise, previous research found that propolis samples from different regions had significantly different inhibitory activity against the growth of *P. larvae* (Bastos et al. 2008, Wilson et al. 2015). Because propolis is a mixture of resins collected by individual bees, it is likely that the great diversity in the ability of samples of propolis to inhibit the growth of *P. larvae* is due to the different resins bees collect from various plant species in different regions (Mihai et al. 2012). Therefore, the next step was to explore whether bees change their foraging preference for specific plant resins after challenge with a bacterial or fungal pathogen.

To test if bees alter their selection of resins after colonies are challenged with a bacterial or fungal pathogen (*P. larvae* and *A. apis*, respectively), resin loads were collected from the hind legs of returning resin foragers during each observation (pre-challenge and post-challenge). Individual resin loads were stored in separate glass vials and the botanical source of the resin was further analyzed in the laboratory.

It is difficult to monitor bees foraging for resin on plants because resin foraging is particularly rare, compared to others types of foraging, and bees often collect resin high in the canopy of trees, which makes it difficult to observe resin foraging directly. The plant source of a resin collected by a bee can be identified by chemically comparing the resin loads of returning foragers with resins collected directly from plants. This strategy is very similar to how pollen foraging is tracked. Since the shapes (morphology) of pollen grains are characteristic of specific plants, microscopy is used to match the morphology of bee-collected pollen to the morphology of pollen collected from flowers. Resins have chemistries that are characteristic of specific plants, and these chemical signatures, rather than morphology, are used to identify resin sources.

In collaboration with M. Wilson, J. Cohen and A. Hegeman from the Horticultural Science Department of the University of Minnesota, resin chemistries were examined using two techniques in series, liquid chromatography and then mass spectrometry (LC-MS). Essentially, LC-MS sorts the hundreds of compounds found in resins by water solubility. This information is then condensed into a "fingerprint." If the chemical pattern, or fingerprint, of a bee-collected resin load is the same as the chemical pattern of a resin collected directly from a plant, it can be concluded that the bee visited that specific plant (Fig. 7).

To date, analysis of data from resin loads collected from bee hind legs in 2012 and 2014 revealed that bees collected resin from five botanical sources in St. Paul, Minnesota: *Populus deltoides* (Eastern cottonwood trees), *P. hybrid* (hybrid poplar trees), and three sources that are not yet identified, unknowns 1, 2 and 3 (Borba 2015). The majority of bees in all colonies collected resin from the most abundant resin-producing tree around the St. Paul campus area, Eastern cottonwood (*P. deltoides*), while resin from the other four sources was not collected in great quantities.

For the most part, all colonies continued to collect resin from the same sources after they were challenged with either the bacterial or fungal pathogen, with the

Fig. 7 Resin fingerprint of Eastern cottonwood trees collected from individual tree buds (**a**), and fingerprint of resin collected from the bee's hind leg (**b**). Based on the similarities of the chemical pattern of these two resin fingerprints, we can conclude that the resin collected from this bee is from an Eastern cottonwood tree

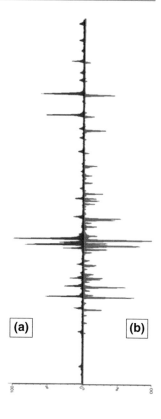

(a)   (b)

exception of colonies in 2012 that did not collect resin from hybrid poplar during the post-challenge period. In general, when colonies increased resin foraging, they simply increased the number of foragers collecting resin from the plants they were already visiting (Borba 2015).

The antimicrobial activity of the resins bees collected was measured in liquid culture using the same assay used to measure the bioactivity of larval food (Sect. 1.2). Of the five different plant sources of resin, the resin from Eastern cottonwood and hybrid poplar had the greatest antimicrobial activity against fungal growth (*A. apis*). Resin from Eastern cottonwood also had the highest antimicrobial activity against bacterial growth (*P. larvae*), but hybrid poplar had relatively low inhibitory activity against this pathogen. Thus, post-challenge colonies did not appear to change their foraging preference to collect resins with higher specific bioactivity (they do not forage for "stronger medicines" for a particular pathogen). Other trees around the St. Paul campus area, such as white spruce (*Picea glauca*), secrete resin with even higher antimicrobial activity against *P. larvae* compared to Eastern cottonwood (Wilson et al. 2013). However, bees apparently do not collect resin from white spruce around the St. Paul campus, as the chemical signatures of the three unknowns did not correspond to white spruce or any other resin-producing plant identified in Wilson et al. (2013).

Bees' decision-making process to collect resin from specific sources after chalk-brood and AFB infection could be driven by the abundance of the plant in the area, the abundance of resin produced by particular plants, the ease of collecting resin from particular plants, distance from the hive, and/or the bioactivity of the resin. Resin collection and choice by bees are unstudied areas that require further investigation.

1.4   Recommendation for Beekeepers

These studies clearly show the benefit of a propolis envelope, particularly an envelope naturally constructed by the bees, to bee health and immune system functioning. The collection of resins to construct a natural propolis envelope is performed by a relatively rare subset of the worker foraging force. The number of resin foragers is probably less than 1% of the total number of foragers in the hive, but this foraging preference may be influenced by the bees' genetics (Butler 1949; Page and Fondrk 1995). Resin collection is partly a genetic tendency and partly a demand-driven process (Martinez and Soares 2012; Nakamura and Seeley 2006). How and what they detect inside the nest to determine need is not clear. When resin foragers encounter rough surfaces and gaps inside the hive, they respond by collecting more resin to seal these cracks in the nest architecture (Simone-Finstrom and Spivak 2010). Therefore, a colony of bees can be encouraged to build a natural propolis envelope within standard beekeeping equipment by modifying the inner walls of bee boxes. Commercial propolis traps can be cut to fit the four inside walls of the hive boxes and stapled with the smooth side of the trap facing the wood and the rough side facing the colony. Using nine frames instead of ten is best when using this method. If the inside of the bee box is built with unfinished, rough lumber, scraped briskly with a wire brush, or if 3 mm grooves are cut in the interior walls of the box, the bees will apply a layer of propolis in the grooves, forming a natural propolis envelope.

A cautionary note for beekeepers. The initial experimental design for the study on the long-term effects of the propolis envelope (see Sect. 1.1) consisted of three treatments: colonies without a propolis envelope (control), colonies with a propolis envelope, and colonies fitted with a propolis trap on top of the frames of the top box, as is done to collect propolis commercially. Bees from colonies with the propolis traps on top of the frames showed inconsistent, and sometimes higher immune-related gene expression, compared to bees in the propolis envelope and control colonies. Moreover, bees from colonies with a propolis trap on top of the frames had significantly higher levels of virus (i.e., DWV) compared to bees in control and propolis envelope treatment colonies in September 2012, May 2013 and May 2014. The presence of high levels of virus has been correlated with colony death and the reduced efficacy of the bee's immune system. It is possible that the presence of the water-resistant propolis trap throughout the year on top of the colony could have altered the microenvironment of the colony (e.g., increasing humidity levels or affecting air circulation within the nest), leading to favorable conditions for the growth of pathogens and maybe viruses. Thus, it appears that leaving a propolis trap on top of a colony for a long period of time, and especially over the winter, is not beneficial to bee health and is not recommended.

Finally, there is no evidence that bees consume resins or propolis. It is not recommended that beekeepers feed propolis solution to bees. Because of the highly antibacterial and antifungal properties of propolis, it could risk killing the beneficial microbiome in bees' guts that is so critical to their health and survival.

1.5 Summary of Findings

Understanding honey bees' natural defense mechanisms allows us to appreciate how resilient honey bees are and to improve our beekeeping practices to enhance their natural behaviors and defenses. The process of domestication of the *Apis mellifera* species by humans using managed hives has interfered with one very important natural defense mechanism of the honey bee colony, the construction of a propolis envelope. The results of research first by M. Simone-Finstrom and later by R. Borba strongly indicate that the propolis envelope serves as an external antimicrobial layer around the colony, providing fundamental benefits to adult bees' immunity (see Sect. 1.1), greater colony fitness in early spring after the winter (see Sect. 1.1), supports bees' natural defense mechanisms against AFB and chalkbrood disease (see Sect. 1.2) and supports nurse bees' ability to induce a strong and effective immune response after AFB infection, resulting in a lower infection load after two months following bacterial challenge (see Sect. 1.2). Honey bees self-medicate by increasing the number of resin foragers after the colony is infected with the fungal pathogen that causes chalkbrood, but not after infection with the bacterial pathogen that causes AFB (see Sect. 1.3). After bacterial or fungal challenge, colonies do not appear to change their resin foraging preference; instead, it appears that bees simply increase resin foraging for resin sources previously collected by the colony. The decision-making process for the recruitment of specific resin sources after chalk-brood and AFB infection, and whether the decisions are driven by plant resin source abundance or resin bioactivity, requires further investigation.

Given all the evidence provided here, it is important to recognize the significance of the propolis envelope as a crucial component of the nest architecture in honey bee colonies. When searching for an apiary location, beekeepers should take into consideration both flower abundance and diversity, and the presence of resin-producing plants within foraging distance from the apiary.

References

Alfonsus EC (1933) Some sources of propolis. Glean Bee Cult 61:92–103
Amdam GV, Norberg K, Hagen A, Omholt SW (2003) Social exploitation of vitellogenin. Proc Natl Acad Sci 100:1799–1802

Amdam GV, Hartfelder K, Norberg K, Hagen A, Omholt SW (2004) Altered physiology in worker honey bees (Hymenoptera: Apidae) infested with the mite *Varroa destructor* (Acari: Varroidae): a factor in colony loss during overwintering? J Econ Entomol 97:741–747

Armbruster SW (1984) The role of resin in Angiosperms pollination: ecological and chemical considerations. Am J Bot 71:1149–1160

Bastos EMAF, Simone M, Jorge DM, Soares AEE, Spivak M (2008) *In vitro* study of the antimicrobial activity of Brazilian propolis against *Paenibacillus larvae*. J Invertebr Pathol 97:273–281

Bilikova K, Wu GS, Simuth J (2001) Isolation of a peptide fraction from honeybee royal jelly as a potential antifoulbrood factor. Apidologie 32:275–283

Bilikova K, Popova M, Trusheva B, Bankova V (2013) New anti-Paenibacillus larvae substances purified from propolis. Apidologie 44:278–285

Bitondi MMG, Simoes ZLP (1996) The relationship between level of pollen in the diet, vitellogenin and juvenile hormone titres in Africanized *Apis mellifera* workers. J Apic Res 35:27–36

Boecking O, Spivak M (1999) Behavioral defenses of honey bees against Varroa jacobsoni Oud. Apidologie 30(2–3):141–158, Springer Verlag (Germany)

Borba RS (2015) Constitutive and therapeutic benefits of plant resins and a propolis envelope to honey bee, *Apis mellifera* L., immunity and health. Doctoral dissertation, University of Minnesota, Minneapolis, MN. (ProQuest/UMI 3734812)

Borba RS, Klyczek KK, Mogen KL, Spivak M (2015) Seasonal benefits of a natural propolis envelope to honey bee immunity and colony health. J Exp Biol. doi:10.1242/jeb.127324

Butler C (1949) The honeybee: an introduction to her sense-physiology and behaviour. Oxford University Press, London

Cantwell GE (1970) Standard methods for counting *Nosema* spores. Am Bee J 110:222–223

Chan QWT, Melathopoulos AP, Pernal SF, Foster LJ (2009) The innate immune and systemic response in honey bees to a bacterial pathogen, *Paenibacillus larvae*. BMC Genomics 10:387

Clayton DH, Wolfe D (1993) The adaptive significance of self-medication. Trends Ecol Evol 8:60–63

Cremer S, Armitage SAO, Schmid-Hempel P (2007) Social immunity. Curr Biol 17:693–702

Engel MS (1999) The taxonomy of recent and fossil honey bees (Hymenoptera: Apidae; Apis). J Hymenopt Res 8:165–196

Engels W, Kaatz H, Zillikens A, Simões ZLP, Trube A, Braun R, Dittrich F (1990) Honey bee reproduction: vitellogenin and caste-specific regulation of fertility. In: Hoshi M, Yamashita O (eds) Advances in invertebrate reproduction. Elsevier, Amsterdam, pp 495–502

Evans JD (2004) Transcriptional immune responses by honey bee larvae during invasion by the bacterial pathogen, *Paenibacillus larvae*. J Invertebr Pathol 85:105–111

Evans JD, Pettis JS (2005) Colony-level impacts of immune responsiveness in honey bees, *Apis mellifera*. Evolution 59:2270–2274

Evans JD, Spivak M (2010) Socialized medicine: Individual and communal disease barriers in honey bees. J Invertebr Pathol 103:S62–S72

Fearnley J (2001) Bee propolis: natural healing from the hive. Souvenir Press, London

Kumazawa S, Yoneda M, Shibata I, Kanaeda J, Hamasaka T, Nakayama T (2003) Direct evidence for the plant origin of Brazilian propolis by the observation of honeybee behavior and phytochemical analysis. Chem Pharm Bull 51:740–742

Langenheim JH (2003) Plant resins: chemistry, evolution, ecology, and ethnobotany. Timber Press, Portland, OR

Lee KV, Moon RD, Burkness EC, Hutchison WD, Spivak M (2010) Practical sampling plans for *Varroa destructor* (Acari: Varroidae) in *Apis mellifera* (Hymenoptera: Apidae) colonies and apiaries. J Econ Entomol 103:1039–1050

Martinez OA, Soares AEE (2012) Melhoramento genético na apicultura comercial para produção da própolis. Rev Bras Saude e Prod Anim 13:982–990

Mattila HR, Otis GW (2006) Influence of pollen diet in spring on development of honey bee (Hymenoptera: Apidae) colonies. J Econ Entomol 99:604–613

Meyer W (1956) Propolis, bees and their activities. Bee World 37:25–36

Mihai CM, Mârghitaş LA, Dezmirean DS, Chirilâ F, Moritz RFA, Schlüns H (2012) Interactions among flavonoids of propolis affect antibacterial activity against the honeybee pathogen *Paenibacillus larvae*. J Invertebr Pathol 110:68–72

Moret Y, Schmid-Hempel P (2000) Survival for immunity: the price of immune system activation for bumblebee workers. Science 290:1166–1168

Nakamura J, Seeley TD (2006) The functional organization of resin work in honeybee colonies. Behav Ecol Sociobiol 60:339–349

Nasr ME, Thorp RW, Tyler TL, Briggs DL (1990) Estimating honey bee (Hymenoptera: Apidae) colony strength by a simple method: measuring cluster size. J Econ Entomol 83:748–754

Nicodemo D, De Jong D, Couto RHN, Malheiros EB (2013) Honey bee lines selected for high propolis production also have superior hygienic behavior and increased honey and pollen stores. Genet Mol Res 12:6931–6938

Nicodemo D, Malheiros EB, De Jong D, Couto RHN (2014) Increased brood viability and longer lifespan of honeybees selected for propolis production. Apidologie 45:269–275

Page RE, Fondrk MK (1995) The effects of colony-level selection on the social organization of honey bee (*Apis mellifera* L.) colonies: colony-level components of pollen hoarding. Behav Ecol Sociobiol 36:135–144

Rose RI, Briggs JD (1969) Resistance to American foulbrood in honey bees IX. Effects of honey-bee larval food on the growth and viability of *Bacillus larvae*. J Invertebr Pathol 13:74–80

Rothenbuhler WC, Thompson VC (1956) Resistance to American foulbrood in honey bees. 1. Differential survival of larvae of different genetic lines. J Econ Entomol 49:470–475

Salmela H, Amdam GV, Freitak D (2015) Transfer of immunity from mother to offspring is mediated via egg-yolk protein vitellogenin. PLoS Pathog 11:e1005015

Schmid-Hempel P (2005) Evolutionary ecology of insect immune defenses. Annu Rev Entomol 50:529–551

Schmitzová J et al. (1998) A family of major royal jelly proteins of the honeybee Apis mellifera L. Cell Mol Life Sci 54:1020–1030. doi:10.1007/s000180050229

Seeley TD, Morse RA (1976) The nest of the honey bee (*Apis mellifera* L.). Insectes Soc 23:495–512

Simone M, Evans JD, Spivak M (2009) Resin collection and social immunity in honey bees. Evolution 63:3016–3022

Simone-Finstrom M, Spivak M (2010) Propolis and bee health: the natural history and significance of resin use by honey bees. Apidologie 41:295–311. doi:10.1051/apido/2010016

Simone-Finstrom MD, Spivak M (2012) Increased resin collection after parasite challenge: a case of self-medication in honey bees? PLoS ONE 7:17–21

Spivak M, Reuter GS (2001a) *Varroa destructor* infestation in untreated honey bee (Hymenoptera: Apidae) colonies selected for hygienic behavior. J Econ Entomol 94:326–331

Spivak M, Reuter GS (2001b) Resistance to American foulbrood disease by honey bee colonies *Apis mellifera* bred for hygienic behavior. Apidologie 32:555–565

Sturtevant AP, Revell IL (1953) Reduction of *Bacillus larvae* spores in liquid food of honey bees by action of the honey stopper, and its relation to the development of American foulbrood. J Econ Entomol 46:855–860

Teixeira ÉW, Negri G, Meira RMSA, Message D, Salatino A (2005) Plant origin of green propolis: Bee behavior, plant anatomy and chemistry. Evidence-based Complement. Altern. Med. 2:85–92

Thompson VC, Rothenbuhler WC (1957) American foulbrood in honey bees. II. Differential protection of larvae by adults of different genetic lines. J Econ Entomol 50:731–737

Wilson MB, Spivak M, Hegeman AD, Rendahl A, Cohen JD (2013) Metabolomics reveals the origins of antimicrobial plant resins collected by honey bees. PLoS ONE 8:1–13

Wilson MB, Brinkman D, Spivak M, Gardner G, Cohen JD (2015) Regional variation in composition and antimicrobial activity of US propolis against *Paenibacillus larvae* and *Ascosphaera apis*. J Invertebr Pathol 124:44–50

Wilson-Rich N, Dres ST, Starks PT (2008) The ontogeny of immunity: Development of innate immune strength in the honey bee (*Apis mellifera*). J Insect Physiol 54:1392–1399

Witham TG (1983) Host manipulation of parasites: within plant variation as a defense against rapidly evolving pests. In: Denno RF, McClure MS (eds) Variable Plants and Herbivores in Natural and Managed Systems. Academic Press, New York, pp 15–41

Author Biographies

Dr. Renata S. Borba is a postdoctoral fellow who has recently joined Dr. Steve Pernal's apiculture research team at Beaverlodge Research Farm, Agriculture Agri-food Canada. Renata received her Ph.D. in Entomology in 2015 from the University of Minnesota studying under Dr. Mara Spivak. Previously, she received her B.Sc. in Animal Science from the Universidade Federal of Ceara, Brazil. Renata's doctoral research focused on evaluating: 1) the seasonal benefits of propolis (a bee-produced resinous material) on the health and immunity of honey bees; 2) the role that types of resins play as a defense against two highly infectious brood pathogens, Ascosphaera apis (a fungus causing chalkbrood disease) and Paenibacillus larvae (a bacterium causing American foulbrood disease); and 3) the effects of the propolis "envelope" within the hive as a natural defense against disease.

Michael B. Wilson is a postdoctoral scientist living and working in Minnesota's Twin Cities. He found his fascination of bees and the people who keep them while training under Marla Spivak and Jerry Cohen at the University of Minnesota. He is currently focused on studying how resinous plants impact honey bee health and understanding the mechanism of benefits bees derive from propolis in their nests. He firmly believes that enhancing and leveraging what bees do naturally to prevent disease will lead to more sustainable beekeeping.

Mike, his wife Fern, and their dog Hannah are enthusiastic gardeners, hikers, and beach bums. When Mike is not doing science or exploring the outdoors, he is getting beat up by his students at Ram's Taekwondo in St. Paul, MN. While his heart is still young, his knees are getting old! He will readily confirm that daily doses of kicking kept him sane throughout graduate school.

Marla Spivak is a Distinguished McKnight Professor in Entomology at the University of Minnesota. She obtained her Ph.D. at the University of Kansas under Dr. Orley Taylor in 1989 on the ecology of Africanized honey bees in Costa Rica. From 1989 to 1992, she was a postdoctoral researcher at the Center for Insect Science at the University of Arizona where she became interested in honey bee hygienic behavior and continued that line of work at the University of Minnesota beginning in 1993. She has bred a line of honey bees, the Minnesota Hygienic line, to defend themselves against diseases and parasitic mites. Current studies include the benefits of propolis to honey bees and the effects of agricultural landscapes and pesticides on honey bee and native bee health.

Predicting Both Obvious and Obscure Effects of Pesticides on Bees

Dr. Jonathan G. Lundgren

Abstract

Pesticides are a necessary component of the monoculture-based food production system. The chemical management of pests can affect non-target organisms, including honey bees. Risk assessment is a way to evaluate the cost–benefit of pesticide use to honey bees and involves understanding the exposure routes and hazards posed by each particular pesticide. The effects of insecticides on bees are intuitively recognized, but other types of pesticides can affect honey bees too. Even "inactive" ingredients in a pesticide formulation can pose a risk to bees. Bees encounter pesticides as they forage in the environment through direct exposure to pesticide applications, and through contaminated resources such as pollen, nectar, water, comb, and propolis. Pesticides can affect bees in myriad ways. The toxicity of pesticides is highly context-specific, challenging risk assessments. Mortality is the most commonly measured effect of pesticides on bees but sublethal effects range from developmental problems, reduced reproductive fitness, diminished overwintering capacity, and numerous behavioral issues that may not kill the bee outright, but may kill the hives. The pervasiveness of pesticides in the environment means that bees cannot avoid exposure to numerous chemicals. Selecting for bees that are adapted to agrichemical-intensive landscapes may be a short-term solution, but the dynamic evolution of chemical use may prohibit long-term tolerances. Beekeepers and farmers need to work together to create and promote reduced chemical intensive food production systems. This is the only long-term answer for the survival of honey bees and biodiversity in general.

J.G. Lundgren (✉)
Ecdysis Foundation, Estelline, SD 57234, USA
e-mail: jon.lundgren@bluedasher.farm

© Springer International Publishing AG 2017
R.H. Vreeland and D. Sammataro (eds.), *Beekeeping – From Science to Practice*,
DOI 10.1007/978-3-319-60637-8_3

Broad scale simplification of the landscape accompanied the rise of industrial-scale food production, and a variety of agrichemicals are used to support these monoculture-based systems. Biodiversity provides substantial resistance to the proliferation of pests in a variety of ways, but this diversity is removed from our food production systems in order to maximize short-term production goals. In the absence of biotic resistance to pest proliferation, land managers rely on pesticides to replace the pest management function provided by diverse biological communities. The downside is these pesticides do not solve the causative problem that produced the pest. Within this context, pests become resistant, and more and new pesticides are required to maintain pests at low densities in a system that is designed for them to excel (i.e., the pesticide treadmill). This is the environment into which honey bees and other pollinators have been inserted, and pesticides inherent in these systems affect pollinators in complex ways. Pesticides are not intended to hurt bees or any other beneficial organisms, but they often do; estimating this harm is called risk assessment.

1 Assessing Risk

Defining the non-target organism is a crucial first step for a valid risk assessment (NRC 1983; Suter 2016). First, it is important to select ecologically relevant species on which to conduct risk assessments (Carignan and Villard 2002). It is not feasible to evaluate the risk of every pesticide against every non-target organism in a habitat. For instance, there are 467 beneficial or neutral insect species in South Dakota sunflowers (Bredeson and Lundgren 2015a), 382 in SD corn (Welch and Lundgren 2016), 150+ in eastern South Dakota dung pats (Pecenka and Lundgren in press). To curtail this list, indicator species that represent certain species groups are often selected to make risk assessments more manageable (NRC 1983), and honey bees are often one of these indicator species (Duan et al. 2008; ECFR 2017). Once the species is selected, the physiological status of the organism affects the outcome of a risk assessment. Life stage, time of day, time of year, history of exposure, nutritional status, reproductive status, exposure to other stressors, social caste, etc., all can influence the perception of risk, and so the context of these risk assessments needs to be clearly defined. A pesticide may have little toxicity to a healthy bee in a Petri dish, but be very toxic to a bee that has been exposed to stressful conditions (e.g., a lack of forage, extreme temperatures, infected with disease, and exposure to other pesticides, etc.).

In its simplest form, risk is defined as hazard $\times$ exposure (NRC 1983). "Hazard" is the negative effect that you are measuring. But even the most hazardous chemical poses no risk if an organism is not exposed to it. Conversely, a fairly benign chemical can be toxic if one is exposed to too much. For example, a single sting from a honey bee is relatively harmless, but a whole hive of stings can be lethal; unless of course you are allergic. Dose often makes the poison, but how an organism is exposed also matters. Whether a substance is ingested, breathed, or physically contacted are types of exposure that influence the risk equation (Vandenberg et al. 2012). Also, some chemicals are only toxic at low doses, while others may actually benefit the organism at low doses (Calabrese 2004; Guedes and Cutler 2014). Hazards posed by pesticides might include increased mortality, reduced reproduction, foraging capability, or honey production (discussed at length later in the chapter). Defining potential hazards at the onset of a risk assessment is critical to an accurate perception of the risk involved. The trouble is, we often times cannot predict how a pesticide is going to adversely affect the environment. For example, who could have foreseen that certain herbicides would alter the sexual characteristics of frogs (Hayes et al. 2002)? The hazard is severe, but risk assessment of this hazard could not be evaluated until we observed the effect in nature. Because we cannot predict all of the risks a pesticide poses to the environment, a precautionary principle is often advocated (Kriebel et al. 2001). In this case, the precautionary principle invokes the notion that the effects of pesticides are unknown and their use unnecessarily or prophylactically should be avoided.

Risk is not unique to pesticides. There are costs and benefits to all decisions, and our current sociological values define how much cost within a certain set of circumstances is acceptable. As society's values, or the environment, or even our ability to characterize hazard changes, our perception of an acceptable level of risk is also altered. For example, DDT was deemed fairly safe when its evaluation was simply based on acute mammalian toxicity. Only when we could measure the widespread bioaccumulation of DDT and its metabolites did we alter our decision and recognize that DDT posed an unreasonable risk posed to the environment (Dunlap 2008; Perkins 1982). Likewise, our society often values threatened species at a higher level than common ones (Mace et al. 2008). So the risks posed by a pesticide that inadvertently kills a portion of a bee population (or a lady beetle population, or a fox population, ad infinitum) does not raise actionable concern until that population is diminished to the point where additional mortality becomes untenable. Given these complexities, it is clear that risk assessments require regular re-evaluations to ensure that risks remain acceptable.

2   Types of Pesticides

Pesticides are categorized at the highest level based on which class of organisms they are designed to control (e.g., herbicides, insecticides, acaricides, rodenticides, fungicides). Within these categories, pesticides are further subdivided based on how they kill the pest (their mode of action). There are hundreds of pesticide active ingredients that are currently registered in the US and Europe (Chauzat et al. 2009; Mullin 2015), but these products represent only a handful of modes of action (i.e., they only affect a handful of physiological targets in a pest). This broad classification system is somewhat misleading. Just because an herbicide is designed to kill plants does not mean that its effects on other groups of organisms will be negligible. Indeed we found that herbicides can be toxic to lady beetles at levels far below the label rate (Freydier and Lundgren 2016). Morton et al. (1972) found that the arsenic herbicides Paraquat, methane arsonic acid (MAA), monosodium methanearsonate (MSMA), disodium methanearsonate (DSMA), hexaflurate, and cacodylic acid were all highly toxic to newly emerged honey bee workers (mortality was significant at 10 ppm). Different classes of pesticides can even synergize to enhance the toxicity of an "insecticide." For instance, adding fungicides (piperonyl butoxide, triflumizole, and propiconazole) increased the oral toxicity of neonicotinoid insecticides (acetamiprid and thiacloprid) to honey bees, sometimes by as much as 1100-fold (Iwasa et al. 2004). The effects of many pesticides on bee health have been investigated, but arguably none have drawn more recent attention than neonicotinoid insecticides. This relatively new group of insecticides targets the nervous system of insects (the neonicotinoids are surrogates for the insect neurotransmitter acetylcholinesterase, to be specific), and are highly toxic to bees (sublethal effects have been observed with as little as 1 billionth of a gram per bee). Given their widespread use and implications in honey bee declines, substantial controversy has surrounded these chemicals (Carreck and Ratnieks 2014; Douglas and Tooker 2016). The effects of neonicotinoids on honey bees will be better explained throughout this chapter.

To complicate matters, the risk posed by a pesticide is strongly influenced by the myriad "inactive" ingredients that are included with the product. Inactive ingredients are classified as surfactants, penetrant enhancers, activators, spreaders, stickers, wetting agents, buffers, antifoaming agents, drift retardants, etc. (Mullin et al. 2015). Registered "inactive" ingredients are largely unregulated; the US Environmental Protection Agency (EPA) only requires that adjuvants submit to seven of the 20 short-term avian and mammalian safety tests that active ingredients must address (Mullin et al. 2015). This can be problematic, as greater amounts of "inactive" ingredients are used, they can sometimes have greater impact on the pest or non-target organisms than the active ingredients within a pesticide formulation (Cox and Surgan 2008; Mann and Bidwell 1999; Surgan et al. 2010). The "inactive" ingredients may also temper the effects of the active ingredient on non-target organisms. One example comes with the herbicide Paraquat, which alone significantly reduces fat body cells (called oenocytes) in bees. But when the adjuvant N-acetylcysteine is added to the formulation, the herbicide has fewer deleterious effects on these cells (Cousin et al. 2013). This notwithstanding, most examples reported in the literature discuss synergistic or additive, deleterious effects on non-target organisms of adding adjuvants and pesticides. One penetrant enhancer, called NMP (N-methyl-2-pyrrolidone), has received recent attention for its toxicity to bees. Hundreds of millions of pounds of NMP are applied in the US alone. This chemical has demonstrated negative effects on wildlife (Mullin et al. 2015), and is itself highly toxic to honey bee larvae (Zhu et al. 2014). Some organosilicone surfactants (Dyne-Amic, Silwet, and Syltac) also have negative effects on honey bees at low concentrations, this time affecting bee learning ability (Ciarlo et al. 2011). When one combines the hundreds of potentially active ingredients with the hundreds of potential "inactive" ingredients, the number of assessments required to understand the risks posed by agrichemicals to bees becomes rather staggering, especially when one considers that formulated products are largely unregulated.

3 Honey Bee Exposure to Pesticides

There are two sources of pesticide exposure for bees, environmental and within-hive exposures. Most of the diversity of chemistries in the hive originates from the environment. Also, the beekeeper can sometimes be his own enemy, and a major source of contaminating pesticides within the hive is those pesticides applied to protect the bees from in-hive pests. Indeed, nearly all of the hives tested in one study had coumaphos and fluvalinate, two acaricides used to manage *Varroa destructor* in infested hives (Mullin et al. 2010). A honey bee can have a much different exposure scenario based on its age and social caste. The oldest workers are the first to be exposed to a particular environmental pesticide, as they are the active foragers (Winston 1987). Moreover, these oldest workers are also the ones to remove the dead bees, and so their exposure could be high if there are pesticide-related deaths in the hives. Once an environmental pesticide enters the hive, middle-aged bees are likely the next to be exposed, as they accept nectar and pollen from the returning foragers (Johnson 2010). Finally, the youngest nurse bees are exposed to pesticides as they feed nectar to the developing larvae and the queen, as well as when they manipulate and build pesticide-impregnated comb. All of this is to say that younger workers may have a very different risk equation than older workers. Likewise, larvae developing in a pesticide-contaminated cell may have a different risk equation than any other age guild or caste within the hive. As such, risk assessments for honey bees are much more complicated than are typically necessary for other non-target species, and simply evaluating the toxicity of a pesticide to a random worker bee is insufficient to assess the risk of a pesticide.

Environmental exposure. The reality is that pesticide use in North America and worldwide is continuing to rise, and some level of pesticides pervade most habitats in the soil, water, and plants. Environmental samples are frequently contaminated with pesticides (Ryberg and Gilliom 2015; Toccalino et al. 2014), including plants and water sources frequented by honey bees (Botias et al. 2015; David et al. 2016; Mogren and Lundgren 2016). In the case of insecticides, fewer pounds of insecticides are applied to farms, but the area treated continues to increase (Fausti et al. 2012), and in some cases, the toxicity of insecticides has increased dramatically from earlier chemistries. In the past 10 years, neonicotinoids have become one of the most commonly used insecticides in North America, and are currently applied to nearly 13% of the land surface of the continental United States (Douglas and Tooker 2015). These neonicotinoids are 5000–10,000 times more toxic to honey bees than DDT (Pisa et al. 2015). Fungicide and herbicide application rates also continue to rise (NASS 2017). Glyphosate is currently applied to the majority of row crop acres around the world (Benbrook 2016); the active ingredient of this herbicide has little toxicity to honey bees, but the Roundup Weathermax® formulation has some deleterious effects on bees (Mullin et al. 2010).

Application technology also affects a honey bee's exposure to a pesticide, and these approaches to deploying pesticides can be generally categorized as broadcast and systemic approaches. Aerial sprays pose a threat of direct contact with the bees and thus are presumed to have the greatest impact on bee workers if applied during foraging peaks. Systemic insecticides, those that are applied to the soil or the seed and then transported throughout the treated plant, may reduce direct exposure of bees to the toxin, but this technique does not eliminate bee exposure. In the case of neonicotinoid seed treatments, very little of the active ingredient is taken up by the treated plant, and the remaining 80–98% (Sur and Stork 2003) of the active ingredient is released into the environment, possibly being transported in the soil and water (Long and Krupke 2016; Morrissey et al. 2015). These insecticides can also be inadvertently disseminated into the environment during planting. When insecticide-coated crop seeds are planted, it accompanies a dust that falls into the surrounding habitats. The neonicotinoids associated with this planter dust can adversely affect honey bee hives, especially those within a certain distance of the planted field (Krupke et al. 2012; Sgolastra et al. 2012; Tapparo et al. 2012). These exposures to "dust off" can be catastrophic for a beekeeper, with a loss of nearly 100% of hives (J.G.L., personal observation). Bees flying during planting can be contaminated with high levels of the neonicotinoid; for example, some bees exposed to the dust had 1240 ng of clothianidin per bee (Tapparo et al. 2012). The systemic nature of neonicotinoid allows them to be taken up by untreated sources of bee forage in the environment (Botias et al. 2015; David et al. 2016; Krupke et al. 2012; Long and Krupke 2016; Pecenka and Lundgren 2015). Mogren and Lundgren (2016) found that untreated flowering strips planted to conserve pollinators near organic and conventional cornfields were contaminated with the neonicotinoid clothianidin, and the level of clothianidin found in the bee bread of contaminated hives was strongly and positively correlated with nutritional stress on the bees. The end result of this substantial environmental exposure is that numerous pesticides are returned to the hive.

In-hive exposure. Pesticide contamination of the hive makes this a dangerous place to live for a honey bee. Hundreds of pesticides and their residues have been isolated from wax comb, pollen, or dead bees within bee hives (Frazier et al. 2015; Long and Krupke 2016; Mullin et al. 2010). In one of the most comprehensive examinations of hive contaminants, Mullin et al. (2010) found that all tested Florida and California hives were contaminated with pesticides and their metabolites, with an average of 6.5 pesticides per hive (118 different pesticides were identified in the study). Of these, nearly half the hives were contaminated with the systemic neonicotinoid insecticides. The majority of samples were contaminated with fluvalinate and coumaphos (two acaricides used to combat *Varroa destructor*), chlorpyrifos (an insecticide), and chlorothalonil (a fungicide). More often than not, multiple pesticides were found in each sample tested. Pesticide exposure within the hive is a consistent stressor on hive health. Comb and propolis, pollen, nectar, water, and dead bees can all be a source of pesticide exposure for bees living exclusively within the hive.

Comb and propolis. Many pesticides are lipophilic and can accumulate to high levels within the wax comb in which the bees spend nearly all of their lives. Once a pesticide enters the comb, it diffuses throughout the wax over a matter of weeks. The exact mechanism behind this transference is unknown. Also, the diversity and quantity of the chemistry found in the comb is subsequently correlated with the pesticides found in the nectar stored in the comb (Byrne et al. 2014) and in dead bees found outside of the hive (Mullin et al. 2010). In one study, Wu et al. (2011) found that pesticides in contaminated comb had moved to adjacent pesticide-free comb within 19 days (a typical brood cycle). Once contaminated, pesticides in the comb can persist for long periods of time. For example, some of the common acaricides used in managing *Varroa* mites can persist for years in the comb (Bogdanov 2004). Other pesticides (e.g., imidacloprid) may persist for much less time (Dively et al. 2015). The end result is that a tremendous diversity of pesticides and their residues are found in the wax comb of nearly all bee hives tested. Eighty-seven and 39 pesticides (or their residues) have been recovered from wax samples from active bee hives in North America, with an average of 6–10 pesticides reported (Mullin et al. 2010; Wu et al. 2011). The most common pesticides found were consistently the acaricides (fluvalinate and coumaphos) used in *Varroa* control, but chlorothalonil (a fungicide) and chlorpyrifos (an insecticide) were also found in most wax samples, as were neonicotinoids. Mullin et al. (2010) found up to 39 pesticides in a single wax sample! The quantities of these pesticides in the comb can be staggering; the average amount of fluvalinate and coumaphos found in the combs were around 6700 ppb and 8300 ppb (respectively), and the maximum quantity of these pesticides found in a single wax sample was more than 22,000 ppb (Wu et al. 2011). Methods for detecting various pesticide groups in propolis have been developed (Chen et al. 2009; dos Santos et al. 2008), but more research is needed on applying these methods to hive-collected propolis samples. Coumaphos and chlorpyrifos were found in nearly all propolis samples analyzed in Uruguay (Pérez-Parada et al. 2011). The implications of pesticide-contaminated comb on beeswax that is sold commercially remains a question. If pesticides are volatilized upon burning, does this pose a health hazard? Two overarching conclusions that can be drawn here are that the bees are consistently living in a matrix of pesticide cocktails and that beekeepers that manage the in-hive pests with pesticides can be exposing their bees to relatively high levels of toxin for long periods of time.

Pollen. The major source of protein for worker bees, queens, and developing larvae is pollen. Hives consume large quantities of pollen, especially during reproductive growth phases of the hive. One study showed that hives collect 40 kg of pollen annually (Villa et al. 2000), and complete exposure scenarios given the amount of pollen collected by typical hives are available (Halm et al. 2006; Rortais et al. 2005). Comprehensive evaluations of pesticide contaminants in pollen or bee bread suggest that this is a major source of toxins for the hive (Chauzat et al. 2006; Long and Krupke 2016). Hundreds of pesticides have been found in bee pollen.

Surveys report between 21 and 32 pesticides in a single pollen sample (Long and Krupke 2016; Mullin et al. 2010; Pettis et al. 2013). Pyrethroid insecticides were found in every pollen sample tested, organophosphates were found in 50% of samples, and fungicides were one of the most common pesticides found (Pettis et al. 2013). Pyrethroids were also the most commonly found insecticide in Indiana pollen samples (Long and Krupke 2016). An average of 7–9 pesticides was found per pollen sample in these surveys. The quantity of pesticides in a single sample is also concerning: one survey found a maximum of 29,000 ppb (an average of 4400 ppb per sample) of the fungicide chlorothalonil in the bee's pollen (Pettis et al. 2013). Neonicotinoids are also frequently found in pollen of seed-treated crops (Bredeson and Lundgren 2015b; Byrne et al. 2014; Krupke et al. 2012). Pollens from untreated wildflowers and conservation strips that are embedded in an agricultural matrix also are frequently contaminated with neonicotinoids (Botias et al. 2015; Chauzat et al. 2006; David et al. 2016; Lu et al. 2015). Hives placed in conservation strips adjacent to cornfields collected pollens that had 10 times the honey bee LD_{50} for clothianidin (Mogren and Lundgren 2016). Although these neonicotinoids are frequently encountered in the pollen, they usually are not the dominant pesticide encountered based on the few investigations published.

Nectar. Simple carbohydrates are a source of rapid energy used by workers and other hive members to fuel flight and basic metabolic processes. Although it consists primarily of simple sugars (sucrose, fructose, and glucose), nectars can have a diversity of micronutrients that influence the biology and behavior of floral visitors (Lundgren 2009). Because nectar is derived from phloem contents, any pesticides that are transported in phloem will often be present in the nectar. Indeed, insecticides can be found in flower nectar within a few days of application (Barker et al. 1980) and can persist for days or even months (Byrne et al. 2014; Waller et al. 1984). Numerous insecticides have been found in floral nectar, including dimethoate, trichlorfon, deltamethrin, Schraden, imidacloprid, clothianidin, phosphamidon, and furadan, among many others (Lundgren 2009). One older literature review found that systemic insecticides were found in floral nectar in 71% of 34 published studies (Davis et al. 1988). Uncontaminated nectar from the field that is stored in insecticide-contaminated wax can become contaminated with fairly high doses; this was observed with imidacloprid in citrus nectar (Byrne et al. 2014). Converting nectar to honey does not necessarily reduce the risk of pesticide contamination (Blasco et al. 2003; Chen et al. 2014; Rissato et al. 2007). In one recent study, 70% of Massachusetts honey samples were contaminated with neonicotinoids, with imidacloprid being particularly prevalent (Lu et al. 2015). In addition to the harm these pesticides and residues pose to the hive itself, these contaminants become problematic when marketing the honey due to food safety regulations.

Water. Honey bees require water to survive and cool the hive, and pesticides can contaminate surface and plant-based water sources at levels that may affect bee hives. Many pesticides contaminate environmental sources of surface water, including ponds, rivers, and streams (Eichelberger and Lichtenberg 1971; Martínez et al. 2000; Schwarzenbach et al. 2010). This contamination is related to both the chemistry of the pesticide itself (e.g., its water solubility, adsorption to soil molecules, and stability in the environment) as well as the environment (proximity to the source of a pesticide, soil physical and chemical properties, biological communities within a habitat, etc.) (Arias-Estévez et al. 2008). Nevertheless, surface waters that are visited by honey bees (Butler 1940; Robinson et al. 1984) are prone to contamination with pesticides, and this exposure pathway is particularly pertinent to agricultural areas where water samples often have higher contamination levels. Foraging workers devote part of their lives to water collection, and a specific caste of workers devotes their efforts exclusively to water collection (Robinson et al. 1984). These bees return to the nest with a crop full of water that they share with other members of the hive (Visscher et al. 1996; Woyciechowski 2007). In addition to surface waters, bees also collect water from guttation fluids from insecticide-treated plants, and this may be an exposure pathway whereby systemic insecticides like neonicotinoids can affect bees (Girolami et al. 2009; Hoffman and Castle 2012; Tapparo et al. 2011). Their systemic nature does not preclude these neonicotinoids from contaminating other environmental sources of water (Main et al. 2016; Morrissey et al. 2015), but we do not entirely understand how these contaminants get from cropland to surface waters. Certainly, more risk assessments should focus on water as a relevant exposure pathway for agrichemicals to affect pollinators.

Dead bees. When bees die from pesticide exposure in the hive, the remaining nest-mates may be adversely affected by pesticide residues in the bee corpses. Also, piles of dead bees in front of the hive can sometimes give an indication of an acute pesticide exposure (Frazier et al. 2015). For example, atrazine (herbicide), metolachlor (herbicide), and clothianidin were found in the corpses of bees piled in front of Indiana hives (Krupke et al. 2012). Often, analysis of dead bees reveals this type of multiple pesticide exposure; an average of 2.5 pesticides were found per bee in one study (Mullin et al. 2010). The exposure level of pesticides revealed by these dead bees can be astounding. Bees flying during corn planting were exposed to planter dust with clothianidin. Flying bees were then collected and allowed to die without further exposure. Some workers had up to 640 ng of clothianidin on their bodies (Tapparo et al. 2012). The half-lives of insecticides on and in dead bees are another consideration. In neonicotinoid-contaminated bees, the parent compound is only detectable for a few hours after exposure (Chauzat et al. 2009; Tapparo et al. 2012); this short half-life may explain why neonicotinoids are not always detected on bee corpses following a "dust off" event. Finally, acute pesticide exposure often kills the bees during foraging, and these poisoned bees never return to the hive. For this reason, it can be difficult to rank risk factors leading to hive declines because direct evidence of pesticide mortality is lacking.

4 How Do Pesticides Affect Bees?

The presence of an active egg laying queen in pheromonal control of colony integrity, sufficient ratio of bees to brood to maintain population growth, relatively disease/pest free, and adequate nutrition are principal determinants of a healthy honey bee colony. (Dively et al. 2015)

The health of the hive is an aggregation of lethal and sublethal effects on the various life stages and social castes over time. Thus, acute toxicity of a pesticide can have predictable effects on hive performance. But pesticides also can exact numerous sublethal effects on individual members of the hives, and these many little hammers can combine into substantial hive-level effects on hive survival under the correct circumstances. These combined lethal and sublethal effects also can increase the negative effects that other stressors such as diseases or pests have on hive health. Moreover, these more obscure sublethal effects can operate at very low doses of pesticides. For instance, locomotor activity was significantly reduced when fipronil was administered at doses 600-fold lower than the LD_{50} for this pesticide (Charreton et al. 2015). For these reasons, risk assessments of pesticides against honey bees can be very challenging to conduct and interpret (Mullin et al. 2015).

Honey bees may be more prone to pesticide effects than other insects. Social insects sometimes sacrifice some of their innate immunity and detoxification capabilities in favor of "social immunity"; this is the case with the honey bee. Individual bees have fewer detoxification enzymes to help nullify pesticide contaminants (Claudianos et al. 2006). But behaviors like nest cleaning and inherent aspects of the hive meant to replace this innate immunity against environmental toxicants may be less effective against pesticides. As mentioned above, comb and propolis which have antibiotic characteristics aggregate pesticides rather than reduce their exposure. One hopeful aspect is that different hive genotypes are differentially affected by pesticides (Laurino et al. 2013; Sandrock et al. 2014), which suggests that selection toward living in a matrix of pesticides should be possible, once natural selection has culled pesticide-susceptible hives.

Here, I document some of the lethal and sublethal effects of pesticides that affect hive health. It is important to note that most of these studies were conducted with a fairly narrow focus, and none consider multiple contributing mechanisms or declines based on pesticide acute or chronic toxicity on bee performance. Suffice it to say that many if not most life history parameters of honey bee hives can be affected by pesticides.

Mortality and survival. Mortality is an easily observed and oft-reported experimental endpoint in hazard assessments. Within a population, there is often a wide range of susceptibilities to even the most toxic substances, and rare resistant individuals can survive high doses. This can make determining a dose that kills 100% of a population challenging and has prompted risk assessors to instead report the doses that kill some (50%) or most (90%) of a population. These are called the LD_{50} and LD_{90} values; other values are also sometimes reported (LD_{80}, LD_{99}, etc.). In the cases when the ingested dose cannot be determined in an assay (for example, when the amount of pesticide-contaminated diet ingested cannot be measured), the lethal concentrations (LC values) that the bees are exposed to are used in lieu of the LD value. The duration and frequency of exposure have great bearing on these LD assessments. Most often, risk assessments of pesticides focus on individual bee mortality under very controlled (e.g., isolated in a laboratory) conditions. One study reports the lethality of a range of pesticides relative to the organochlorine insecticide DDT (Pisa et al. 2015). Some newer formulations that have much less active ingredient applied in the environment (like neonicotinoids) likely pose as much if not more hazard to honey bees than some earlier insecticidal chemistries, due to their lowered $LD_{50}s$. Also, a particular pesticide may kill larvae and adults at different rates. For example, Wu et al. (2011) showed that larval mortality was unaffected by comb pesticides, but the longevity of workers was reduced by pesticide exposure. Yet in another study, larvae were much more susceptible to pesticides than the adults, possibly because the larvae only defecate at the end of the stage, prolonging exposure to ingested pesticides (Zhu et al. 2014). One of the most toxic pesticides to bees that I was able to find reported in the literature is Fipronil. Significant mortality was experienced at 0.1 ng/adult bee after a 7 d exposure (Aliouane et al. 2009). At some level, mortality of individual hive members will contribute to hive collapse, but this is a dynamic process that is difficult to predict. The ratio of mortalities inflicted on larvae, workers, and reproductives ultimately combine to form an aggregate risk to the hive itself. Also, contextual considerations like the condition of the hive prior to a pesticide exposure, hive age, or other stressors on the hive all could contribute to the lethality of a pesticide on the hive. Thus, the question of how much mortality is too much mortality is a challenging one to answer.

Development. Alterations in larval development can have cascading negative effects on adult bees and hive population dynamics. Pesticide exposure can slow larval development, and sometimes these effects are seen at doses much lower than LD_{50} values for adult workers (Davis et al. 1988; Wu et al. 2011). Delayed development rates could affect the duration of susceptibility to pests like *Varroa destructor*, as well as decrease the hive growth rate. Hive size is correlated with strength and its ability to survive other stressors. Larval development rates are infrequently reported relative to other hive fitness parameters, but warrant additional attention from researchers.

Mobility and behavior. Many insecticides function in part as neurotoxins, taking advantage of the unique characteristics of insect nervous systems to minimize acute effects on non-insect animals. Some classes of insecticides are surrogates for the insect neurotransmitter enzyme acetylcholine esterase (AChE). Their use overexcites the acetylcholine receptor by replacing the enzyme, causing the nerve cells to continually fire (these compounds do not allow nerves to switch off). This insecticide mode of action is employed by many organophosphates, carbamates, and neonicotinoids (Barker et al. 1980; Boily et al. 2013; Iwasa et al. 2004). Many pyrethroids and organochlorines are also neurotoxins, but affect the ability of nerve cells to repolarize and effect an action potential (these pesticides switch nerves off). Placing hives near neonicotinoid-treated corn fields is sufficient to alter the AChE levels in adult workers (Boily et al. 2013). Targeting these receptors can lead to other effects on the nervous system. Field-relevant doses of imidacloprid can impair the development of mushroom bodies (calyces) in the brain, which are organs with large quantities of AChE receptors (Peng and Yang 2016). These mushroom bodies are where learning occurs, which affects many other aspects of the natural history of honey bees. Impairment of nerve function affects several measurable characteristics of honey bees, including mobility, learning, behavior, orientation, foraging, walking, and communication of the honey bee.

A common measurement of learning capacity is the proboscis extension reflex (PER) to food rewards, and this approach has been used multiple times to demonstrate how pesticides interfere with honey bee learning (El Hassani et al. 2008; Ramirez-Romero et al. 2005). It is also possible that the pesticide active ingredient and its metabolites have different effects on the PER (Guez et al. 2001). At very low doses of imidacloprid (1.5 ng per bee), the PER was increased over the control, but as dose and time went on, the bees became lethargic and unresponsive. One explanation for this may be that the metabolites of the imidacloprid are more effective at reducing learning compared to the parent compound (Lambin et al. 2001). Another explanation for the observation of increased PER response at low doses of neonicotinoids is that nerves over firing requires energy from carbohydrates; as the dose increases, it shuts down the metabolism in the insect. In this way, neurotoxin exposure can manifest itself in altered feeding behavior. Demares et al. (2016) found that thiamethoxam exposure did not affect the PER to protein-based foods, but did alter responsiveness to sucrose solution at certain pesticide and sucrose concentrations. Ability to recognize and imbibe water can also be affected after ingesting pesticides (Aliouane et al. 2009). So, the nutritional status of bees and entire hives can be compromised when learning ability is reduced by a pesticide; mobility and locomotor activity also affects the nutritional stress of the hive by altering foraging behavior.

Short- and long-distance dispersal is affected by pesticides, and sublethal effects on mobility by a pesticide can have important ramifications for the hive (Matsumoto 2013; Ramirez-Romero et al. 2005). Sublethal doses of thiamethoxam and clothianidin (5 and 2 ppb, respectively) reduced foraging success and lowered pollen and nectar collections by treated hives (Sandrock et al. 2014). More specifically, neonicotinoids disrupt navigation capabilities, and the affected bees struggle to find their way back to the nest (Fischer et al. 2014). Pesticide-treated and untreated foraging workers equally found their way to the intended floral resources, but harmonic radar attached to the honey bees revealed that treated workers were significantly less likely to remember the direction back to their nest. This inability to return to the nest appears could be related to navigation rather than on flight capability (Fischer et al. 2014; Matsumoto 2013), both of which are adversely affected by neonicotinoids (Blanken et al. 2015). Walking is another important behavior that can be affected by neurotoxic pesticides. Sublethal doses (10–50-fold lower than the $LD_{50}s$) of pyrethroid and neonicotinoid insecticides reduced walking speeds and distances in adult honey bees (Charreton et al. 2015). Walking may seem trivial to a hive-dwelling insect with flight, but ability to disperse resources throughout the hive, communicate foraging sites (e.g., with the waggle dance), clean the hive, thermoregulate, etc., all depend on locomotor (i.e., walking) behavior.

Winter survival. The aggregate effects of many small detriments to hive health may manifest themselves in the overwintering success of a honey bee hive. In one study, sublethal doses of neonicotinoid insecticides were administered to hives and their performance was compared to untreated hives. Summer performance was equivalent in the two groups, but significantly fewer treated hives survived the winter; the lack of dead bees in failed hives was reminiscent of colony collapse disorder (Lu et al. 2014). This winter mortality resulting from sublethal exposures to imidacloprid follows a standard dose-response curve (Dively et al. 2015). The stress of overwintering then is the final blow to a pesticide-weakened hive.

Reproduction. Queen and drone health dictate the hive growth rate. Several studies have demonstrated direct physiological effects of pesticides on the physiology and fecundity of honey bee queens. Ovarial and spermathecal development was reduced in queens that had been reared on as little as 1 ppb of clothianidin (or 4 ppb of thiamethoxam). Queen survival was reduced by 25%, the number of sterile females were increased and the proportion of queens laying fertilized eggs was reduced by 38%. Sperm viability, number of spermatozoa, and ovariole size were also adversely affected by the neonicotinoids (Williams et al. 2015). Drones are also adversely affected by low levels of neonicotinoids (1.5 ppb clothianidin or 4.5 ppb of thiamethoxam), where adult drone longevity and sperm viability were both reduced significantly (Straub et al. 2016). Lethal and sublethal effects can combine into outright queen failure and supersedure within the hive. For example, imidacloprid administered as 20 and 100 ppb prompted queen failure in late summer, following a broodless period (Dively et al. 2015). Similarly, field-relevant doses of clothianidin and thiamethoxam administered to larvae over two brood cycles resulted in reduced brood production, and greater queen supersedure rates in the treated hives (Sandrock et al. 2014). These effects on reproductive capacity of the hive are not restricted to neonicotinoids. The miticides fluvalinate and coumaphos increase queen mortality and coumaphos lowered queen body weight, reduced ovary size, and lowered the number of sperm in exposed relative to untreated queens (Haarmann et al. 2002).

Susceptibility to other stressors. Pesticides and other stressors interact, sometimes in unpredictable ways. As a result, pesticide toxicity is not always well correlated with a specific response variable in a narrowly focused experimental design without considering other contributing factors to experimental outcomes. Pesticides, diseases and pests are often synergistic or additive in their effects on the toxicity of pesticides to honey bees. *Nosema ceranae* spore counts and impact on the hive are aggravated when bees are simultaneously exposed to one of several pesticides in the diet (Dively et al. 2015; Pettis et al. 2012, 2013) or pesticide-contaminated comb (Wu et al. 2012). Combined deleterious effects of these two stressors accrete over time, and even low doses (e.g., one-hundredth of the LD_{50}) of a pesticide can significantly increase its lethality when combined with *N. ceranae* infection (Retschnig et al. 2014; Vidau et al. 2011). This is in part because the pesticide lowers the innate and social immune responses of the honey bees, making them more susceptible to pathogen infection (Alaux et al. 2010). One behavioral response that has been observed is that bees exposed to neonicotinoids and *N. ceranae* consume more sugar resources, which then exposes them to additional pesticide (Alaux et al. 2010; Vidau et al. 2011).

Varroa mite infestations and pesticide exposures can combine to reduce hive performance more than the individual stressors. When combined, imidacloprid and *Varroa* mites reduced the flight distance and flight time of affected honey bees more than either stressor did alone (Blanken et al. 2015). The hives that died in spring flew significantly shorter distances than the surviving hives, suggesting that fitness reductions produced by both imidacloprid and *Varroa* mite effects may reduce hive survival. These effects may also be the result of suppressed anti-viral immunity in bees that were exposed to low doses (one thousandth of the LD_{50}) of pesticides (Di Prisco et al. 2013). Deformed wing virus, a pathogen vectored by *Varroa* mites, replicated significantly more following the host's exposure to imidacloprid and clothianidin (but not chlorpyrifos).

5   Conclusions

If pesticides were the answer to pest problems, then we should have overcome pests decades ago. Pesticides are not the sole cause of bee declines; they are an artifact of a simplified agroecosystem. Siloing the pesticide issue will not solve the bee problem. The most effective way to reduce the impact of pesticides is to reform our food production systems in which the bees must live. Diversifying our food production system will have knock-on effects like providing additional and diversified forage that bees use to reduce nutritional stress. By improving honey bee nutrition and reducing their toxin exposure, bee immune function and resistance to pests will be promoted. Anything less than reforming this food production system will not solve the problem of bee declines. Paradigm shifts of this nature are not entirely within the beekeepers' ability to control, but there are things that beekeepers and others can do.

6   Suggestions for Beekeepers

From a research and regulatory standpoint, risk assessments of pesticides need to be conducted within a realistic context. The risk is context dependent, and risk assessments need to factor in the many stressors that affect a hive's performance when assessments are conducted. Moreover, hives are systems, and toxicological assessments on hive components (e.g., brood or worker survival) outside of the context of the hive system do not give a true perception of risk. Test scenarios that simultaneously account for numerous lethal and sublethal effects of pesticide exposure over time on hive performance may help to overcome doubt and ambiguity regarding the importance of pesticides in international bee declines. The current infrastructure that funds science fosters doubt regarding the role that pesticides play in bee declines. Beekeepers need to fight fire with fire and fund the independent science and scientists that are willing to pursue these often career-altering research projects into the truth underlying pesticides and bees. The money will not likely come from sources other than the beekeepers (individually and as club groups).

A research priority for beekeepers should be finding non-chemical alternatives for *Varroa* mite control. Beekeepers need to understand that their use of acaricides is compromising the integrity and longevity of their hives. Moreover, the acaricides persist in the hives, so decisions to apply could have long-term implications for the contamination of treated hardware. For these reasons, prophylactic applications of pesticides to control *Varroa* should be avoided. Also, hygienic bee lines (those bees that clean themselves of mites), organic acids, essential oils, and natural enemies of the *Varroa* may be acceptable alternatives to pesticides that could replace the stressors of pesticides in the hive.

The use of pesticides continues to rise, and food production systems are not going to reform overnight. By feeding and treating bee hives, beekeepers slow the adaptation of their hives to living within a pesticide-contaminated landscape matrix. Preventing short-term hive losses with interventions may well be fostering a long-term extension of bee declines. Balancing an operation's profitability with selecting for pesticide-tolerant bee genetics is a central challenge facing the bee industry and hobbyists alike.

Pesticides are ubiquitous in the environment and in the hives, and can be highly toxic to bees; what can a beekeeper do? Most farmers do not understand that there is a better way to farm than conventional, high-input monoculture systems. The nature of beekeeping is such that beekeepers often know many of the farmers in their communities. They know the farmers that are farming ecologically and in diversified systems, and they know those who are not. Worldwide bee declines have initiated tremendous media attention, and beekeepers are frequently the stars of this attention. Beekeepers need to make the ecologically based farmers in their communities into heroes. Take the media attention that has been given and turns the stories about "the bee problem" into one about "the solution," which must come from the farmers themselves.

References

Alaux C, Brunet JL, Dussaubat C, Mondet F, Tchamitchan S, Cousin M, Brillard J, Baldy A, Belzunces LP, Le Conte Y (2010) Interactions between *Nosema* microspores and a neonicotinoid weaken honeybees (*Apis mellifera*). Environ Microbiol 12:774–782

Aliouane Y, El Hassani AK, Gary V, Armengaud C, Lambin M, Gauthier M (2009) Subchronic exposure of honeybees to sublethal doses of pesticides: effects on behavior. Environ Toxicol Chem 28:113–122

Arias-Estévez M, López-Periago E, Martínez-Carballo E, Simal-Gándara J, Mejuto J-C, García-Río L (2008) The mobility and degradation of pesticides in soils and the pollution of groundwater resources. Agr Ecosyst Environ 123:247–260

Barker RJ, Lehner Y, Kunzmann MR (1980) Pesticides and honey bees: nectar and pollen contamination in alfalfa treated with dimethoate. Arch Environ Contam Toxicol 9:125–133

Benbrook CM (2016) Trends in glyphosate herbicide use in the United States and globally. Environ Sci Europe 28:3

Blanken LJ, van Langevelde F, van Dooremalen C (2015) Interaction between *Varroa destructor* and imidacloprid reduces flight capactiy of honeybees. Proc Royal Soc B 282:20151738

Blasco C, Fernández M, Pena A, Lino C, Silveira MI, Font G, Picó Y (2003) Assessment of pesticide residues in honey samples from Portugal and Spain. J Agric Food Chem 51:8132–8138

Bogdanov S (2004) Beeswax: quality issues today. Bee World 85:46–50

Boily M, Sarrasin B, DeBlois C, Aras P, Chagnon M (2013) Acetylcholinesterase in honey bees (*Apis mellifera*) exposed to neonicotinoids, atrazine, and glyphosate: laboratory and field experiments. Environ Sci Pollut Res 20:5603–5614

Botias C, David A, Horwood J, Abdul-Sada A, Nicholls E, Hill E, Goulson D (2015) Neonicotinoid residues in wildflowers, a potential route of chronic exposure for bees. Environ Sci Technol 49:12731–12740

Bredeson MM, Lundgren JG (2015a) Foliar and soil arthropod communities of sunflower (*Helianthus annuus*) fields of central and eastern South Dakota. J Kansas Entomol Soc 88: 305–315

Bredeson MM, Lundgren JG (2015b) Thiamethoxam seed treatments have no impact on pest numbers or yield in cultivated sunflowers. J Econ Entomol 108:2665–2671

Butler CG (1940) The choice of drinking water by the honeybee. J Exp Biol 17:253–261

Byrne FJ, Visscher PK, Leimkuehler B, Fischer D, Grafton-Cardwell EE, Morse JG (2014) Determination of exposure levels of honey bees foraging on flowers of mature citrus trees previously treated with imidacloprid. Pest Manag Sci 70:470–482

Calabrese EJ (2004) Hormesis: a revolution in toxicology, risk assessment and medicine. EMBO Rep 5:S37–S40

Carignan V, Villard M-A (2002) Selecting indicator species to monitor ecological integrity: a review. Environ Monit Assess 78:45–61

Carreck NL, Ratnieks FLW (2014) The dose makes the poison: have "field realistic" rates of exposure of bees to neonicotinoid insecticides been overestimated in laboratory studies? J Apic Res 53:607–614

Charreton M, Decourtye A, Henry M, Rodet G, Sandoz J-C, Charnet P, Collet C (2015) A locomotor deficit induced by sublethal doses of pyrethroid and neonicotinoid insecticides in the honeybee *Apis mellifera*. PLoS ONE 10:e0144879

Chauzat M-P, Carpentier P, Martel A-C, Bougeard S, Cougoule N, Porta P, Lachaize J, Madec F, Aubert M, Faucon J-P (2009) Influence of pesticide residues on honey bee (Hymenoptera: Apidae) colony health in France. Environment Entomology 38:514–523

Chauzat M-P, Faucon J-P, Martel A-C, Lachaize J, Cougoule N, Aubert M (2006) A survey of pesticide residues in pollen loads collected by honeybees in France. J Econ Entomol 99:253–262

Chen F, Chen L, Wang Q, Zhou J, Xue X, Zhao J (2009) Determination of organochlorine pesticides in propolis by gas chromatography–electron capture detection using double column series solid-phase extraction. Anal Bioanal Chem 393:1073–1079

Chen M, Tao L, McLean J, Lu C (2014) Quantitative analysis of neonicotinoid insecticide residues in foods: implications for dietary exposures. J Agric Food Chem 62:6082–6090

Ciarlo TJ, Mullin CA, Frazier JL, Schmehl DR (2011) Learning impairment in honey bees caused by agricultural spray adjuvants. PLoS ONE 7:e40848

Claudianos C, Ranson H, Johnson RM, Biswas S, Schuler MA, Berenbaum MR, Feyereisen R, Oakeshott JG (2006) A deficit of detoxification enzymes: pesticide sensitivity and environmental response in the honeybee. Insect Mol Biol 15:615–636

Cousin M, Silva-Zacarin E, Kretzschmar A, El Maataoui M, Brunet J-L, Belzunces LP (2013) Size changes in honey bee larvae oenocytes induced by exposure to Paraquat at very low concentrations. PLoS ONE 8:e65693

Cox C, Surgan M (2008) Unidentified inert ingredients in pesticides: implications for human and environmental health. Environ Health Perspect 114:1803–1806

David A, Botias C, Abdul-Sada A, Nicholls E, Rotheray EL, Hill EM, Goulson D (2016) Widespread contamination of wildflower and bee-collected pollen with complex mixtures of neonicotinoids and fungicides commonly applied to crops. Environ Int 88:169–178

Davis AR, Solomon KR, Shuel RW (1988) Laboratory studies on honeybee larval growth and development as affected by systemic insecticides at adult-sublethal levels. J Apic Res 27: 146–161

Démares FJ, Crous KL, Pirk CWW, Nicolson SW, Human H (2016) Sucrose sensitivity of honey bees is differently affected by dietary protein and a neonicotinoid pesticide. PLoS ONE 11: e0156584

Di Prisco G, Cavaliere V, Annoscia D et al (2013) Neonicotinoid clothianidin adversely affects insect immunity and promotes replication of a viral pathogen in honeybees. Proc Natl Acad Sci USA 110:18466–18471

Dively GP, Embrey MS, Kamel A, Hawthorne DJ, Pettis JS (2015) Assessment of chronic sublethal effects of imidacloprid on honey bee colony health. PLoS ONE 10:e0118748

dos Santos TFS, Aquino A, Dórea HS, Navickiene S (2008) MSPD procedure for determining buprofezin, tetradifon, vinclozolin, and bifenthrin residues in propolis by gas chromotography-mass spectrometry. Anal Bioanal Chem 390:1425–1430

Douglas MR, Tooker JF (2015) Large-scale deployment of seed treatments has driven rapid increase in use of neonicotinoid insecticides and preemptive pest management in U.S. field crops. Environ Sci Technol 49:5088–5097

Douglas MR, Tooker JF (2016) Meta-analysis reveals that seed-applied neonicotinoids and pyrethroids have similar negative effects on abundance of arthropod natural enemies. Peer J 4: e2776

Duan JJ, Marvier M, Huesing JE, Dively GP, Huang Z (2008) A meta-analysis of effects of Bt crops on honey bees (Hymenoptera: Apidae). PLoS ONE 3:e1415

Dunlap TR (ed) (2008) DDT, *silent spring*, and the rise of environmentalism. University of Washington Press, Seattle, pp 1–150

ECFR (2017) Electronic code of federal regulations: data requirements for pesticides, Vol. web-based document. United States Government Publishing Office

Eichelberger JW, Lichtenberg JJ (1971) Persistence of pesticides in river water. Environ Sci Technol 5:541–544

El Hassani AK, Dacher M, Gary V, Lambin M, Gauthier L, Armengaud C (2008) Effects of sublethal doses of acetamiprid and thiamethoxam on the behavior of the honeybee (*Apis mellifera*). Arch Environ Contam Toxicol 54:653–661

Fausti SW, McDonald TM, Lundgren JG, Li J, Keating AR, Catangui MA (2012) Insecticide use and crop selection in regions with high GM adoption rates. Renewable Agric Food Syst 27:295–304

Fischer J, Müller A, Spatz AK, Greggors U, Grünewald B, Menzel R (2014) Neonicotinoids interfere with specific components of navigation in honeybees. PLoS ONE 9:e91364

Frazier MT, Mullin CA, Frazier JL, Ashcraft SA, Leslie TW, Mussen EC, Drummond FA (2015) Assessing honey bee (Hymenoptera: Apidae) foraging populations and the potential impact of pesticides on eight U.S. crops. J Econom Entom 108:2141–2152

Freydier L, Lundgren JG (2016) Unintended effects of the herbicides 2,4-D and dicamba on lady beetles. Ecotoxicology 25:1270–1277

Girolami V, Mazzon L, Squartini A, Mori N, Marzaro M, Di Bernardo A, Greatti M, Giorio C, Tapparo A (2009) Translocation of neonicotinoid insecticides from coated seeds to seedling guttation drops: A novel way of intoxication for bees. J Econ Entomol 102:1808–1815

Guedes RNC, Cutler GC (2014) Insecticide-induced hormesis and arthropod pest management. Pest Manag Sci 70:690–697

Guez D, Suchail S, Gauthier M, Maleszka R, Belzunces LP (2001) Contrasting effects of imidacloprid on habituation in 7 and 8 day old honeybees (*Apis mellifera*). Neurobiol Learn Mem 76:183–191

Haarmann T, Spivak M, Weaver D, Weaver B, Glenn T (2002) Effects of fluvalinate and coumaphos on queen honey bees (Hymenoptera: Apidae) in two commercial queen rearing operations. J Econ Entomol 95:28–35

Halm M-P, Rortais A, Arnold G, Tasei JN, Rault S (2006) New risk assessment approach for systemic insecticides: the case of honeybees and imidacloprid (Gaucho). Environ Sci Technol 40:2448–2454

Hayes T, Haston K, Tsui M, Hoang A, Haeffele C, Vonk A (2002) Herbicides: feminization of male frogs in the wild. Nature 419:895–896

Hoffman EJ, Castle SJ (2012) Imidacloprid in melon guttation fluid: a potential mode of exposure for pest and beneficial organisms. J Econ Entomol 105:67–71

Iwasa T, Motoyama N, Ambrose JT, Roe RM (2004) Mechanism for the differential toxicity of neonicotinoid insecticides in the honey bee, *Apis mellifera*. Crop Protection 23:371–378

Johnson BR (2010) Division of labor in honeybees: form, function, and proximate mechanisms. Behav Ecol Sociobiol 64:305–316

Kriebel D, Tickner J, Epstein P, Lemons J, Levins R, Loechler EL, Quinn M, Rudel R, Schettler T, Stoto M (2001) The precautionary principle in environmental science. Environ Health Perspect 109:871–876

Krupke CH, Hunt GJ, Eitzer BD, Andino G, Given K (2012) Multiple routes of pesticide exposure for honey bees living near agricultural fields. PLoS ONE 7:e29268

Lambin M, Armengaud C, Raymond S, Gauthier M (2001) Imidacloprid induced facilitation of the proboscis extension reflex habituation in the honeybee. Arch Insect Biochem Physiol 48:129–134

Laurino D, Manino A, Patetta A, Porporato M (2013) Toxicity of neonicotinoid insecticides on different honey bee genotypes. Bull Insectol 68:119–126

Long EY, Krupke CH (2016) Non-cultivated plants present a season-long route of pesticide exposure for honey bees. Nature Commun 7:11629

Lu C, Chang C-H, Chen M (2015) Distributions of neonicotinoid insecticides in the Commonwealth of Massachusetts: a temporal and spatial variation analysis for pollen and honey samples. Environ Chem 13:4–11

Lu C, Warchol KM, Callahan RA (2014) Sub-lethal exposure to neonicotinoids impaired honey bees winterization before proceeding to colony collapse disorder. Bull Insectol 67:125–130

Lundgren JG (2009) Relationships of natural enemies and non-prey foods. Springer International, Dordrecht, The Netherlands

Mace GM, Collar NJ, Gaston KJ, Hilton-Taylor C, Akçakaya HR, Leader-Williams N, Milner-Gulland EJ, Stuart SN (2008) Quantification of extinction risk: IUCN's system for classifying threatened species. Conserv Biol 22:1424–1442

Main AR, Michel NL, Cavallaro MC, Headley JV, Peru KM, Morrissey CA (2016) Snowmelt transport of neonicotinoid insecticides to Canadian Prairie wetlands. Agr Ecosyst Environ 215:76–84

Mann RM, Bidwell JR (1999) The toxicity of glyphosate and several glyphosate formulations to four species of southwestern Australian frogs. Arch Environ Contam Toxicol 36:193–199

Martínez RC, Gonzalo ER, Laespada MEF, San Román FJS (2000) Evaluation of surface- and ground-water pollution due to herbicides in agricultural areas of Zamora and Salamanca (Spain). Journal of Chromatography A 869:471–480

Matsumoto T (2013) Reduction in homing flights in the honeybee *Apis mellifera* after a sublethal dose of neonicotinoid insecticides. Bull Insect 66:1–9

Mogren CL, Lundgren JG (2016) Neonicotinoid-contaminated pollinator strips adjacent to cropland reduce honey bee nutritional status. Sci Rep 6:29608

Morrissey CA, Mineau P, Devries JH, Sanchez-Bayo F, Liess M, Cavallaro MC, Liber K (2015) Neonicotinoid contamination of global surface waters and associated risk to aquatic invertebrates: a review. Environ Int 74:291–303

Morton HL, Moffett JO, MacDonald RH (1972) Toxicity of herbicides to newly emerged honey bees. Environ Entomol 1:102–104

Mullin CA (2015) Effects of 'inactive' ingredients on bees. Curr Opin Insect Sci 10:194–200

Mullin CA, Chen J, Fine JD, Frazier MT, Frazier JL (2015) The formulation makes the honey bee poison. Pest Biochem Physiol 120:27–35

Mullin CA, Frazier M, Frazier JL, Ashcraft S, Simonds R, VanEngelsdorp D, Pettis JS (2010) High levels of miticides and agrochemicals in North American apiairies: implications for honey bee health. PLoS ONE 5:e9754

NASS (2017) National Agriculture Statistics Service. USDA

National Research Council (1983) Risk assessment in the federal government: managing the process. National Academy Press, Washington, D.C.

Pecenka JR, Lundgren JG (2015) Non-target effects of clothianidin on monarch butterflies. Sci Nat 102:19

Pecenka JR, Lundgren JG (in press) Contributions of herd management to dung arthropod communities in eastern South Dakota. J Med Entomol

Peng Y-C, Yang E-C (2016) Sublethal dosage of imidacloprid reduces the microglomerular density of honey bee mushroom bodies. Sci Rep 6:19298

Pérez-Parada A, Colazzo M, Besil N, Geis-Asteggiante L, Rey F, Heinzen H (2011) Determination of coumaphos, chlorpyrifos and ethion residues in propolis tinctures by matrix solid-phase dispersion and gas chromatography coupled to flame photometric and mass spectrometric detection. J Chromatogr A 1218:5852–5857

Perkins JH (1982) Insects, experts, and the insecticide crisis. Plenum Press, New York

Pettis JS, Lichtenberg EM, Andree M, Stitzinger J, Rose R, vanEngelsdorp D (2013) Crop pollination exposes honey bees to pesticides which alters their susceptibility to the gut pathogen *Nosema ceranae*. PLoS ONE 8:e70182

Pettis JS, vanEngelsdorp D, Johnson J, Dively G (2012) Pesticide exposure in honey bees results in increased levels of the gut pathogen *Nosema*. Naturwissenschaften 99:153–158

Pisa LW, Amaral-Rogers V, Belzunces LP, Bonmatin JM, Downs CA, Goulson D, Kreutzweiser DP, Krupke C, Liess M, McField M, Morrissey CA, Noome DA, Settele J, Simon-Delso N, Stark JD, Van der Sluijs JP, Van Dyck H, Wiemers M (2015) Effects of neonicotinoids and fipronil on non-target invertebrates. Environ Sci Pollut Res 22:68–102

Ramirez-Romero R, Chaufaux J, Pham-Delègue M-H (2005) Effects of Cry1Ab protoxin, deltamethrin and imidacloprid on the foraging activity and the learning performances of the honeybee *Apis mellifera,* a comparative approach. Apidologie 36:601–611

Retschnig G, Neumann P, Williams GR (2014) Thiacloprid-*Nosema ceranae* interactions in honey bees: host survivorship but not parasite reproduction is dependent on pesticide dose. J Invertebr Pathol 118:18–19

Rissato SR, Galhiane MS, de Almeida MV, Gerenutti M, Apon BM (2007) Multiresidue determination of pesticides in honey samples by gas chromatography-mass spectrometry and application in environmental contamination. Food Chem 101:1719–1726

Robinson GE, Underwood BA, Henderson CE (1984) A highly specialized water-collecting honey bee. Apidologie 15:355–358

Rortais A, Arnold G, Halm M-P, Touffet-Briens F (2005) Modes of honeybees exposure to systemic insecticides: estimated amounts of contaminated pollen and nectar consumed by different categories of bees. Apidologie 36:71–83

Ryberg KR, Gilliom RJ (2015) Trends in pesticide concentrations and use for major rivers of the United States. Sci Total Environ 538:431–444

Sandrock C, Tanadini M, Tanadini LG, Fauser-Misslin A, Potts SG, Neumann P (2014) Impact of chronic neonicotinoid exposure on honeybee colony performance and queen supersedure. PLoS ONE 9:e103592

Schwarzenbach RP, Egli T, Hofstetter TB, von Gunten U, Wehrli B (2010) Global water pollution and human health. Annu Rev Environ Resour 35:109–136

Sgolastra F, Renzi T, Draghetti S, Medrzycki P, Lodesani M, Maini S, Porrini C (2012) Effects of neonicotinoid dust from maize seed-dressing on honeybees. Bull Insectology 65:273–280

Straub L, Villamar-Bouza L, Bruckner S, Chantawannakul P, Gauthier L, Khongphinitbunjong K, Retschnig G, Troxler A, Vidondo B, Neumann P, Williams GR (2016) Neonicotinoid insecticides can serve as inadvertent insect contraceptives. Proc R Soc B 283:20160506

Sur R, Stork A (2003) Uptake, translocation and metabolism of imidacloprid in plants. Bull Insect 56:35–40

Surgan M, Condon M, Cox C (2010) Pesticide risk indicators: unidentified inert ingredients compromise their integrity and utility. Environ Manage 45:834–841

Suter GWI (2016) Ecological risk assessment, 2nd edn. CRC Press, Boca Raton, FL

Tapparo A, Giorio C, Marzaro M, Marton D, Solda L, Girolami V (2011) Rapid analysis of neonicotinoid insecticides in guttation drops of corn seedlings obtained from coated seeds. J Environ Monit 13:1564–1568

Tapparo A, Marton D, Giorio C, Zanella A, Soldá L, Marzaro M, Vivan L, Girolami V (2012) Assessment of the environmental exposure of honeybees to particulate matter containing neonicotinoid insecticides coming from corn coated seeds. Environ Sci Technol 46:2592–2599

Toccalino PL, Gilliom RJ, Lindsey BD, Rupert MG (2014) Pesticides in groundwater of the United States: decadal-scale changes, 1993–2011. Groundwater 52:112–125

Vandenberg LN, Colborn T, Hayes TB et al (2012) Hormones and endocrine-disrupting chemicals: low-dose effects and nonmonotonic dose responses. Endocr Rev 33:378–455

Vidau C, Diogon M, Aufauvre J, Fontbonne R, Viguès B, Brunet JL, Texier C, Biron DG, Blot N, El Alaoui H, Belzunces LP, Delbac F (2011) Exposure to sublethal doses of fipronil and thiacloprid highly increases mortality of honeybees previously infected by *Nosema ceranae*. PLoS ONE 6:e21550

Villa S, Vighi M, Finizio A, Serini GB (2000) Risk assessment for honeybees from pesticide-exposed pollen. Ecotoxicology 9:287–297

Visscher PK, Crailsheim K, Sherman G (1996) How do honey bees (*Apis mellifera*) fuel their water foraging flights? J Insect Physiol 42:1089–1094

Waller GD, Erickson BJ, Harvey J, Martin JH (1984) Effects of dimethoate on honey bees (Hymenoptera: Apidae) when applied to flowering lemons. J Econ Entomol 77:70–74

Welch KD, Lundgren JG (2016) An exposure-based, ecology-driven framework for selection of indicator species for insecticide risk assessment. Food Webs 9:46–54

Williams GR, Troxler A, Retschnig G, Roth K, Yanez O, Shutler D, Neumann P, Gauthier L (2015) Neonicotinoid pesticides severely affect honey bee queens. Sci Rep 5:14621

Winston ML (1987) The biology of the honey bee. Harvard University Press, Boston, MA

Woyciechowski M (2007) Risk of water collecting in honeybee (*Apis mellifera*) workers (Hymenoptera: Apidae). Sociobiology 50:1–10

Wu JY, Anelli CM, Sheppard WS (2011) Sub-lethal effects of pesticide residues in brood comb on worker honey bee (*Apis mellifera*) development and longevity. PLoS ONE 6:e14720

Wu JY, Smart MD, Anelli CM, Sheppard WS (2012) Honey bees (*Apis mellifera*) reared in brood combs containing high levels of pesticide residues exhibit increased susceptibility to *Nosema* (Microsporidia) infection. J Invertebr Pathol 109:326–329

Zhu W, Schmehl DR, Mullin CA, Frazier JL (2014) Four common pesticides, their mixtures and a formulation solvent in the hive environment have high oral toxicity to honey bee larvae. PLoS ONE 9:e77547

Author Biography

Dr. Jonathan G. Lundgren is an agroecologist, Director of ECDYSIS Foundation, and CEO for Blue Dasher Farm. He received his Ph.D. in Entomology from the University of Illinois in 2004 and was a top scientist with USDA-ARS for 11 years. Lundgren received the Presidential Early Career Award for Science and Engineering by the White House. Lundgren has served as an advisor for national grant panels and regulatory agencies on pesticide and GM crop risk assessments. Lundgren has written 107 peer-reviewed journal articles, authored the book "Relationships of Natural Enemies and Non-prey Foods," and has received more than $3.4 million in grants. He has trained 5 post-docs and 12 graduate students from around the world. One of his priorities is to make science applicable to end users, and he regularly interacts with the public and farmers regarding pest and farm management and insect biology. Lundgren's research and education programs focus on assessing the ecological risk of pest management strategies and developing long-term solutions for sustainable food systems. His ecological research focuses heavily on conserving healthy biological communities within agroecosystems by reducing disturbance and increasing biodiversity within cropland.

Sublethal Effects of Pesticides on Queen-Rearing Success

Gloria DeGrandi-Hoffman and Yanping Chen

Abstract

The effects of sublethal pesticide exposure on queen emergence and immunity were measured. Queen-rearing colonies were fed pollen with chlorpyrifos (CPF) alone (pollen-1) and with CPF and the fungicide Pristine® (pollen-2). Fewer queens emerged when larvae were reared in colonies fed pollen-1 or pollen-2 than when larvae were reared in outside colonies without contaminated pollen. Larvae grafted from and reared in colonies fed pollen-2 had the lowest rate of queen emergence. Deformed wing virus (DWV) and black queen cell virus were found in nurse bees from colonies fed pollen-1 or pollen-2 and in outside colonies. The viruses also were detected in queen larvae. However, we did not detect virus in emerged queens grafted from and reared in outside colonies. In contrast, DWV was found in all emerged queens grafted from colonies fed pollen-1 or pollen-2 and reared either in outside hives or those fed pollen-1 or pollen-2. The results suggest that sublethal exposure of CPF alone but especially when Pristine® is added reduces queen emergence possibly due to compromised immunity in developing queens.

There are many reasons for colony losses: disease, Varroa, poor nutrition, and lethal exposure to pesticides. In recent years, one of the more common causes of colony losses is queen failure. Queens are being replaced at unprecedented rates in managed colonies. In some cases, when queens are lost, the colony is unable to rear a

G. DeGrandi-Hoffman (✉)
Carl Hayden Bee Research Center, USDA-ARS, 2000 East Allen Road,
Tucson, AZ 85719, USA
e-mail: Gloria.Hoffman@ARS.USDA.GOV

Y. Chen
USDA-ARS Bee Research Laboratory, 10300 Baltimore Ave., Bldg. 306,
Rm. 315, BARC-EAST, Beltsville, MD 20705, USA
e-mail: Judy.Chen@ARS.USDA.GOV

© Springer International Publishing AG 2017
R.H. Vreeland and D. Sammataro (eds.), *Beekeeping – From Science to Practice*,
DOI 10.1007/978-3-319-60637-8_4

new queen to emergence. If the beekeeper does not introduce a new queen, the colony will die.

Though queenless colonies might not be able to replace their queen, a new queen can be purchased and introduced in the colony. However, queen breeders in California reported that they were unable to rear queens when they fed Queen-rearing colonies pollen collected from almond orchards. Sublethal levels of pesticides and fungicides that are often present in almond pollen were suspected of being a contributing cause.

To determine whether sublethal exposure to pesticides affected Queen-rearing, two of the most common contaminants in pollen were tested for their effects on queen emergence. The compounds were chlorpyrifos (CPF) often applied as Dursban® or Lorsban® and the fungicide Pristine® (PRS), that is, a combination of Boscalid and Pyraclostrobin (Mullin et al. 2010). CPF is applied before bloom in fruit and nut orchards to control scale insects. PRS can be applied during bloom to stone fruit (e.g., almonds, cherries, apricots, plums, and peaches) and pome fruit trees (apples and pears) to prevent brown rot, blossom blight, powdery mildew, scab, leaf spot, and shot hole, and to strawberries to prevent Botrytis gray mold, leaf spot, and powdery mildew (http://agproducts.basf.us/products/pristine-fungicide.html).

CPF is an organophosphate (OP) insecticide that affects the nervous system by inhibiting neurotransmitters (Pope 1999). However, there is increasing evidence that OPs have other biological effects (Duysen et al. 2001; Pettis et al. 2012, 2013). For example, OPs may disrupt metabolism (Adigun et al. 2010) and alter immune function by oxidative stress and subsequent tissue damage or stress-related immunosuppression (Li 2007).

Unlike insecticides that might target neural function, fungicides can affect basic cellular processes such as nucleic acid and protein synthesis, the structure and function of cell membranes, mitosis and cell division (Yang et al. 2011). The fungicide used in this study (PRS) affects the production of ATP, a molecule that supplies large amounts of energy for biochemical processes in cells. ATP is produced in specialized cellular structures called mitochondria. Fungicides that compromise ATP production in fungi can also reduce ATP levels in non-target organisms that synthesize this molecule in their mitochondria. Recently, PRS was reported to lower ATP levels in honey bees and reduce protein digestion (DeGrandi-Hoffman et al. 2015). There is increasing evidence that compounds that reduce ATP production also affect immune responses because these require the energy that ATP provides (Arnoult et al. 2009; Wu et al. 2012).

If sublethal exposure to pesticides affects immunity among immature and adult bees, colony losses attributed to viruses or other pathogens or parasites actually could be downstream effects of pesticides. High pathogen titers might also affect the ability of worker bees to rear new queens. In our study, we examined the effects that feeding colonies pollen contaminated with CPF alone and with added PRS might have on rearing queens to emergence. Virus titers in the colonies also were measured to determine whether consuming pollen with these pesticides was affecting immunity.

1   Experimental Procedures and Findings

To provide a greater understanding of how the effects of fungicides on queen development were determined, a brief description of the methods used in this study is provided. For a full description of all methods, see DeGrandi-Hoffman et al. (2013). Pollen was collected using pollen traps on hives placed in almond orchards for pollination. A pesticide analysis revealed the presence of CPF in the pollen.

The pollen collected in the traps was ground to a fine powder and spread on large aluminum trays. We applied PRS to half of the ground pollen at a rate of 3000 ppb of active ingredients (boscalid = 1966 ppb and pyraclostrobin = 998 ppb) using a hand sprayer (Fig. 1). The remaining half of the pollen was sprayed only with distilled water. By treating the pollen this way, we could feed the same source pollen to colonies and compare the effects of CPF alone (pollen-1) and with added PRS (pollen-2) on Queen-rearing success. We also measured the effects on immunity by measuring virus titers in the queens and the nurse bees that reared them.

Fig. 1 a Applying either distilled water (pollen-1) or fungicide (pollen-2) to almond pollen contaminated with chlorpyrifos. **b** Colonies were placed in an enclosed flight area that has 10 separate sections that are 1.93 m wide, 8.25 m long, and 4.14 m high. **c** The sections are separated by cloth mesh and bees cannot fly between the sections. **d** The pollen-1 or pollen-2 was placed at the entrances of colonies in the enclosed flight area. **e** The bees collected the pollen, and **f** used it to rear queens we grafted with larvae from outside free foraging colonies or in colonies fed pollen-1 or pollen-2

2 Effects of Ingesting Contaminated Pollen on Queen-Rearing Success

All experiments were conducted at the USDA-ARS Carl Hayden Bee Research Center, Tucson, AZ. Colonies were comprised of Italian bees (*Apis mellifera ligustica*) and headed by commercially produced and mated European queens (Koehnen and Sons Inc., Glenn, CA, USA). The hives were established in an enclosed flight area in order to control their food source (Fig. 1). These hives were considered as 'enclosed colonies.' All colonies were comprised of about 3000 adult bees and a laying queen. The hives contained frames with foundation, and bees were fed sugar syrup to draw comb. When we saw larvae in the drawn comb, we began feeding the ground pollen to the colonies. There was no pollen in the colonies prior to our pollen feeding. Each colony was fed either pollen-1 or pollen-2. The ground pollen was placed at the hive entrance daily (Fig. 1), and the bees readily collected and stored it. Four colonies were provided with pollen-1 and five with pollen-2. After 4 weeks of feeding on the pollen, we began Queen-rearing experiments in these hives. This procedure insured that the queen cells evaluated in the study were tended by nurse bees reared entirely on the pollen we fed.

We established a second set of 5-frame nucleus hives in an apiary adjacent to the Bee Center. The colonies contained 3000–4000 bees with 2–3 frames of brood. The bees open foraged on native desert vegetation. These colonies are referred to as 'outside hives.'

Larvae (<36 h old) were grafted from worker cells into queen cups and reared into queens using the procedures described in Laidlaw (1979). Colonies used for Queen-rearing were made queenless for 24 h before queen cups containing larvae were introduced. Ten queen cups with larvae were placed in the center of each colony, and combs with bees and brood were placed on either side. The cells remained in the colony until queens were within 48 h of emergence. At that time, the cells were removed and placed in individual sterile vials in an environmental room with a temperature of 32–34°C (approximately 89–92°F) and 50% humidity. Each vial had a small piece of queen candy (a mixture of powdered sugar, honey, and water formed into a paste) for the emerged queen to feed on while in the vial. The queen cells were checked daily for emerged queens.

In the first experiment, larvae were grafted from the outside hives and placed in either enclosed colonies fed pollen-1 or pollen-2 or in different outside hives from those where they were grafted (Fig. 2). In Experiment 2, larvae were grafted into queen cups from the enclosed hives fed either pollen-1 or pollen-2. The queen cups were placed either back in the same enclosed colony from which they were grafted or in outside colonies (Fig. 3). The outside colonies were different from those used in the first experiment.

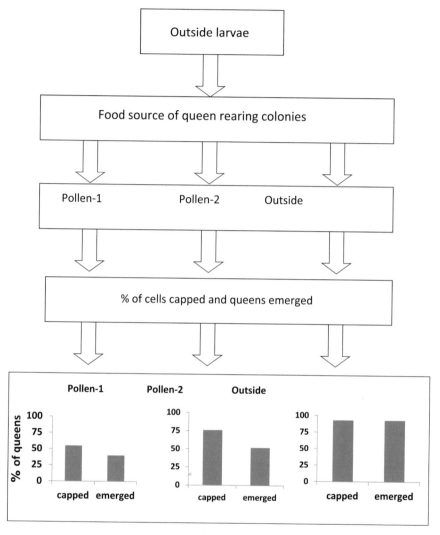

Fig. 2 Percentages of queen cells that were capped and queens that emerged when larvae from open foraging colonies outside an enclosed flight area were reared as queens in colonies fed pollen with chlorpyrifos alone (Pollen-1) or with added Pristine® fungicide (Pollen-2). Based on Chi-square (X^2) tests, significantly more queen cells were capped and had emerging queens when reared in colonies foraging on outside pollen compared with pollen-1 (55% survival to capped stage; $X^2 = 12.3$, $p < 0.0001$, and 40% emergence; $X^2 = 21.0$, $p = 0.0001$) or pollen-2 (76% survival to capped stage $X^2 = 4.03$, $p = 0.046$, and 51.8% emergence; $X^2 = 13.1$, $p < 0.0001$). More larvae survived to the capped brood stage in colonies fed pollen-2 than pollen-1, but the percentage of queens that emerged did not differ (pollen-1 = 40%, pollen-2 = 51.8%; $X^2 = 1.29$, $p = 0.255$)

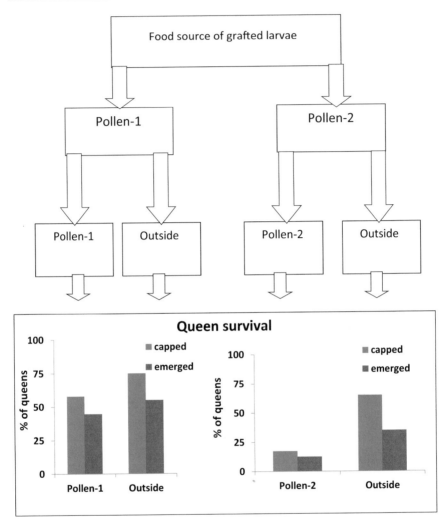

Fig. 3 Percentage of queen cells that were capped and had emerged queens when larvae were grafted from colonies in an enclosed flight area (EFA) that were fed pollen with chlorpyrifos alone (Pollen-1), added Pristine® fungicide (Pollen-2), or were in open foraging colonies outside the EFA (outside). Based on Chi-square (X^2) tests, percentages of larvae grafted from colonies fed pollen-1 that survived to the capped stage did not differ between those reared in colonies fed pollen-1 (57.9%) or in outside colonies (75%) ($X^2 = 1.65$, $p = 0.198$). The percent that emerged also did not differ ($X^2 = 0.55$, $p = 0.457$). Significantly, fewer larvae from colonies fed pollen-2 and reared in those colonies survived to the capped stage ($X^2 = 13.54$, $p < 0.0001$) and emerged as queens compared with those reared in outside colonies ($X^2 = 4.21$, $p = 0.04$). The percentage of larvae grafted from and reared in colonies fed pollen-1 that survived to the capped stage or emerged as queens was significantly higher than when larvae were grafted from and reared in colonies fed pollen-2 (capped stage: $X^2 = 13.6$, $p < 0.0001$, emerged queens: $X^2 = 10.0$, $p = 0.002$)

When larvae were grafted from and reared in outside colonies (Experiment 1), 93% emerged as queens (Fig. 2). This was a significantly higher emergence success than when the larvae were reared in colonies fed either pollen-1 or pollen-2. Only about 40% of the larvae emerged as queens in colonies fed pollen-1, and about 50% emerged in colonies fed pollen-2. In Experiment 2, the fewest larvae emerged as queens when they were grafted from and reared in colonies fed pollen-2 (Fig. 3). Less than 25% of the queen cells survived to the capped stage and even fewer emerged. Larvae grafted from colonies fed pollen-2 and reared in outside colonies also had relatively low queen emergence.

When larvae were grafted from colonies fed pollen-1, the percentages that survived to the capped stage were similar between those reared in colonies fed pollen-1 (57.9%) and in outside colonies (75%). The percentages of larvae grafted from and reared in colonies fed pollen-1 that survived to the capped brood stage or emerged as queens were significantly higher than when larvae were grafted from and reared in colonies fed pollen-2. When queen cells that did not emerge in both Experiments 1 and 2 were opened, we found that either the larvae did not successfully pupate and were a black viscous mass in the cell or were fully formed dark black pupa but were dead in the cells (Fig. 4).

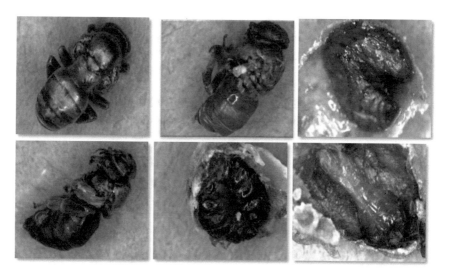

Fig. 4 Examples of queens that failed to emerge when reared in colonies fed almond pollen with either chlorpyrifos alone or with chlorpyrifos and the fungicide Pristine®. Queens that died prior to pupation or were fully formed and pigmented but did not emerge occurred in colonies fed either type of pollen

3   The Effects of Ingesting Contaminated Pollen on Virus Levels

To determine whether ingesting contaminated pollen affects immunity in developing queens, virus titers were measured in: (1) nurse bees tending queen cells, (2) queen larvae developing in the cells, and (3) emerged queens. The samples for virus analyses were collected from colonies in Experiments 1 and 2.

RNA extraction techniques (Chen et al. 2005) were used to test for the presence and relative quantity of 7 common bee viruses including *acute bee paralysis virus* (ABPV), *black queen cell virus* (BQCV), *chronic bee paralysis virus* (CBPV), *deformed wing virus* (DWV), *Kashmir bee virus* (KBV), *Israeli acute paralysis virus* (IAPV), and *Sacbrood virus* (SBV). We consistently found only *deformed wing virus* (DWV) and *Black Queen Cell Virus* (BQCV) in the samples. Detection of the virus does not indicate that the adult bee or larvae showed symptoms of viral disease.

In Experiment 1, DWV was detected in all nurse bees tending the queen cells in both outside colonies and in the enclosed colonies fed pollen-1 or pollen-2 (Fig. 5). About 70% of the queen larvae reared in outside colonies and all of those reared in enclosed colonies had DWV. We did not detect DWV in emerged queens reared in outside colonies. However, about 30% of the emerged queens grafted from larvae in colonies fed pollen-1 and 75% of those from colonies fed pollen-2 had DWV.

BQCV was found more frequently in nurse bees from colonies fed pollen-2 than in nurses from outside colonies or those fed pollen-1. The virus was not detected in queen larvae reared in colonies fed pollen-1 or pollen-2 or in emerged queens. These results might have occurred because those with BQCV died shortly after being grafted and were removed by the bees before we sampled them.

In Experiment 2, we detected DWV in all nurse bees, queen larvae, and virgin queens in both outside and enclosed colonies. BQCV was detected in 83% of the nurse bees in outside colonies and in all of those in the enclosed colonies. More than half of the queen larvae reared in outside colonies and all of those reared in the enclosed colonies had BQCV. All virgin queens grafted from enclosed colonies fed pollen-1 and half of those grafted from colonies fed pollen-2 and reared in outside colonies had BQCV. The virus also was detected in 67% of the virgin queens grafted from and reared in enclosed colonies fed pollen-1 and 33% of those fed pollen-2.

(a) Larvae grafted from enclosed colonies and reared in outside colonies

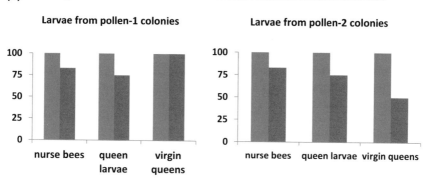

(b)   Larvae grafted from colonies fed pollen-1 or pollen-2 and reared in those colonies

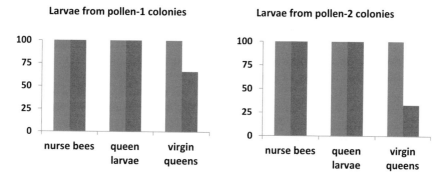

Fig. 5 Percentage of samples testing positive for deformed wing virus (■) and black queen cell virus (■) when larvae were grafted from outside colonies and reared into queens in hives fed almond pollen with chlorpyrifos (pollen 1) or with chlorpyrifos + Pristine® fungicide (pollen 2) (**a**), or were grafted from and reared in colonies fed either pollen-1 or pollen-2 (**b**)

4 Conclusions

Larvae either grafted from or reared as queens in colonies fed pollen contaminated with chlorpyrifos alone or with the fungicide Pristine® were less likely to emerge and more likely to test positive for both DWV and BQCV than those grafted from and reared in colonies without contaminated pollen. Though chlorpyrifos alone appears to reduce queen emergence, the reductions were greater when the fungicide also was present.

The differences in queen emergence rates between larvae reared in outside colonies and enclosed colonies fed contaminated pollen indicate that both the larvae and the Queen-rearing environment were affected by the pesticides. The highest queen emergence occurred when larvae were grafted from and reared in the outside colonies. We did not detect virus in these queens. However, when larvae from the outside colonies were reared in hives fed either pollen-1 or pollen-2, only about 50% of them emerged and DWV was detected in all of them. The difference in queen emergence and virus titers between rearing environments might be partially attributed to greater stress experienced by enclosed colonies compared with those outside where bees were open foraging. Though bees foraged in the enclosed area, the confinement might have produced stress that was not experienced by the outside colonies. Still, the pesticides in the pollen seemed to have an effect on Queen-rearing that was greater than differences in the location of the Queen-rearing colonies. For instance, when larvae were grafted from enclosed colonies and reared in the outside colonies, they had lower emergence rates and tested positive for virus with greater frequency than when larvae were grafted from and reared in outside colonies. Though the larvae selected for grafting were less than 36 h old, the effects of the pesticides were already present.

The findings from this study suggest that there could be severe reductions in queen cell capping and emergence if colonies contain pollen contaminated with pesticides. Colonies that have lost their queen or that are established from queenless splits should be checked frequently to be sure that a new queen is reared to emergence. If beekeepers see cells that were started and then torn down or are sealed but queens do not emerge, the pollen might be too contaminated for the bees to successfully rear queens. Similarly, queen producers should not use pollen in their Queen-rearing colonies that might be contaminated with pesticides and fungicides even if the worker bees seem to be unaffected. Beekeepers that rear queens in colonies with contaminated pollen should be aware that the queens that do emerge might have DWV and thus extend the impact of pollen contamination into the colonies headed by those queens. Though the effects of DWV on the queens themselves seem to be minimal, there is an association between DWV infection in the ovaries and the degeneration of individual follicles (Gauthier et al. 2011). More importantly, DWV is transmitted vertically from queens to their off-spring (Chen et al. 2005; Yue et al. 2006; DeMiranda and Fries 2008) resulting in a persistent latent infection circulating in the colony population. Under appropriate environmental or biological stressors such as Varroa mites, the viruses become activated and cause various pathologies in the hosts. These include behavioral deficiencies (Iqbal and Mueller 2007), wing deformity, and significantly reduced life expectancy (DeMiranda and Genersch 2010). There also is a strong association between DWV in worker bees and colony mortality over winter (Gauthier et al. 2011; Highland et al. 2009; Berthoud et al. 2010; Genersch et al. 2010; DiPrisco et al. 2011). Thus, the sublethal effects of pesticides might be contributing to the pervasive presence of DWV in managed colonies and to colony losses when the viruses are activated by stress factors such as Varroa and the miticide treatments used to control this parasite (Locke et al. 2012).

Though CPF alone affected queen emergence rates and virus titers, the impact was greater when combined with PRS. The combination of a neurotoxin (CPF) and an inhibitor of mitochondrial function (PRS) might have caused the care of the queen cells by nurse bees to be compromised so that the queen larvae were nutritionally stressed. This alone could have reduced the number of capped cells and emerged queens. In addition, CPF and PRS also could have been affecting innate immunity in the nurse bees so that they were transmitting higher virus titers to the larvae. Indeed, there was higher mortality prior to the capped cell stage when larvae were grafted from and reared in colonies containing pollen contaminated with CPF and PRS. Thus, the effects of CPF that is a common contaminant of pollen might be amplified if it is present when PRS is applied to crops in bloom.

Large pollen stores collected from orchards are tempting to use for Queen-rearing colonies or for queenless splits. However, the pollen might be contaminated with pesticides particularly fungicides applied during bloom. To be sure that queens can be reared using the pollen source, beekeepers should graft just a few bars of queen cells to determine whether the queens emerge. If they do, the pollen probably is fine to use for Queen-rearing. However, if very few queen cells are sealed or emerge, the pollen might contain levels of pesticides that are affecting queen development. Determining the effects of the pollen source on queen emergence on a small scale by grafting a few bars could save time in the end if few queens emerge.

References

Adigun AA, Seidler FJ, Slotkin TA (2010) Disparate developmental neurotoxicants converge on the cyclic AMP signaling cascade revealed by transcriptional profiles *in vitro* and *in vivo*. Brain Res 1316:1–16

Arnoult D, Carneiro L, Tattoli I, Girardin SE (2009) The role of mitochondria in cellular defense against microbial infection. Semin Immunol 21:223–232

Berthoud H, Imdorf A, Haueter M, Radloff S, Neumann P (2010) Virus infections and winter losses of honey bee colonies (*Apis mellifera*). J Apic Res 49:60–65

Chen YP, Pettis JS, Feldlaufer MF (2005) Detection of multiple viruses in queens of the honey bee, *Apis mellifera* L. J Invert Pathol 90:118–121

De Miranda JR, Fries I (2008) Venereal and vertical transmission of deformed wing virus in honeybees (*Apis mellifera* L.). J Invertebr Pathol 98:184–189

De Miranda JR, Genersch E (2010) Deformed wing virus. J Invertebr Pathol 103:48–61

DeGrandi-Hoffman G, Chen Y, Simonds R (2013) The effects of pesticides on queen rearing and virus titers in honey bees (*Apis mellifera* L.). Insects 4:71–89, doi:10.3390/insects4010071

DeGrandi-Hoffman G, Chen Y, Watkins deJong E et al (2015) Effects of oral exposure to fungicides on honey bee nutrition and virus levels. J Econ Entomol 108:2518–2528, doi:10.1093/jee/tov251

DiPrisco G, Zhang X, Pennacchio F et al (2011) Viral dynamics of persistent and acute virus infections in honey bee. Viruses 3:2425–2441

Duysen EG, Li B, Xie W, Schopfer LM, Anderson RS, Broomfield CA, Lockridge O (2001) Evidence for nonacetylcholinesterase targets of organophosphorus nerve agent: Supersensitivity of acetylcholinesterase knockout mouse to VX lethality. J Pharmacol Exp Ther 299:528–535

Gauthier L, Ravallec M, Tournaire M, Cousserans F, Bergoin M, Dainat B, deMiranda JR (2011) Viruses associated with ovarian degeneration in *Apis mellifera* L. queens. PLoS ONE 6:e16217

Genersch E, von der Ohe W, Kaatz H et al (2010) The German bee monitoring project: A long term study to understand periodically high winter losses of honey bee colonies. Apidologie 41:332–352

Highland AC, El Nagar A, Mackinder LCM et al (2009) Deformed wing virus implicated in overwintering honeybee colony losses. Appl Environ Microbiol 75:7212–7220

Iqbal J, Mueller U (2007) Virus infection causes specific learning deficits in honeybee foragers. Proc Roy Soc Lond B Biol Sci 274:1517–1521

Laidlaw HH (1979) Contemporary queen rearing. Dadant & Sons, Hamilton, IL, USA

Li Q (2007) New mechanism of organophosphorous pesticide-induced immunotoxicity. J Nippon Med Sch 74:92–105

Locke B, Forsgren E, Fries I, deMiranda JR (2012) Acaricide treatment affects viral dynamics in *Varroa destructor*-infested honey bee colonies via both host physiology and mite control. Appl Environ Microbiol 78:227–235

Mullin CA, Frazier M, Frazier JL, Ashcraft S, Simonds R, vanEngelsdorp D, Pettis JS (2010) High levels of miticides and agrochemicals in North American apiaries: Implications for honey bee health. PLoS ONE 5:e975

Pettis JS, vanEngelsdorp D, Johnson J, Dively G (2012) Pesticide exposure in honey bees results in increased levels of the gut pathogen Nosema. Naturwissenschaften 99:153–158

Pettis JS, Lichtenberg EM, Andree M, Stitzinger J, Rose R, vanEngelsdorp D (2013) Crop pollination exposes honey bees to pesticides which alters their susceptibility to the gut pathogen *Nosema ceranae*. PLoS ONE 8(7):e70182. doi:10.1371/journal.pone.0070182

Pope CN (1999) Organophosphorus pesticides: they all have the same mechanism of toxicity? J Toxicol Environ Health Part B Crit Rev 2:161–181

Wu JY, Smart MD, Anelli CM, Sheppard WS (2012) Honey bees (*Apis mellifera*) reared in brood combs containing high levels of pesticide residues exhibit increased susceptibility to Nosema (Microsporidia) infection. J Invertebr Pathol 109:326–329

Yang C, Hamel C, Vujanovic V, Gan Y (2011) Fungicide: modes of action and possible impact on nontarget microorganisms. ISRN Ecol. doi:10.5402/2011/130289

Yue C, Schroder M, Bienefeld K, Genersch E (2006) Detection of viral sequences in semen of honeybees (*Apis mellifera*): evidence for vertical transmission of viruses through drones. J Invertebr Pathol 92:105–108

Author Biography

Dr. Gloria DeGrandi-Hoffman is the Research Leader at the USDA-ARS Carl Hayden Bee Research Center in Tucson Arizona. She received her B.S. in biology and M.S. in entomology at Penn State University. She received her Ph.D. in entomology from Michigan State. She specializes in mathematical modeling of populations, but also conducts research in honey bee nutrition, Varroa mites, and the effects of sublethal pesticide exposure on honey bee colony health. During her research career, she has conducted studies on the pollination of fruit trees and row crops, Africanized honey bees, and the process of queen replacement in European and African colonies.

Fungi and the Effects of Fungicides on the Honey Bee Colony

Jay A. Yoder, Blake W. Nelson, Andrew J. Jajack and Diana Sammataro

Abstract
Fungicides are found in agricultural areas to protect crops from fungal diseases. When sprayed in areas where honey bee colonies are placed for pollination, bees can collect or otherwise be exposed to these compounds. While labeled safe for bees by the manufacturers, our research found that pollen containing fungicides had a negative effect on the beneficial fungi found in bee colonies that help converting the pollen into bee bread and can end up in the bees themselves. As a result, pathogenic fungi (that cause chalkbrood and stonebrood diseases) were not kept in check by the beneficial fungi, including *Aspergillus*, *Penicillium*, *Cladosporium*, and *Rhizopus*, which were compromised by the presence of fungicides in the hive. Colonies were found to be weakened by the persistent presence of fungicides. Steps to help protect colonies are outlined.

J.A. Yoder (✉) · B.W. Nelson
Department of Biology, Wittenberg University, Springfield, OH 45501, USA
e-mail: jyoder@wittenberg.edu

B.W. Nelson
e-mail: nelsonb@wittenberg.edu

A.J. Jajack
Department of Biomedical, Chemical and Environmental Engineering,
University of Cincinnati, Cincinnati, OH 45267, USA
e-mail: jajackaj@mail.uc.edu

D. Sammataro
DianaBrand Honey Bee Research Services, LLC, Tucson, AZ 85718, USA
e-mail: dsammbeegirl@gmail.com

D. Sammataro
Carl Hayden Bee Research Center, United States Department of Agriculture,
Agricultural Research Service, Tucson, AZ 85719, USA

© Springer International Publishing AG 2017
R.H. Vreeland and D. Sammataro (eds.), *Beekeeping – From Science to Practice*,
DOI 10.1007/978-3-319-60637-8_5

1 Background

Fungicides are routinely used for broad-scale treatment of agricultural and commercial crops to prevent plant diseases (Fig. 1), particularly in orchards such as almonds, apples, peaches (Fig. 2). The problem is that bees forage for pollen and nectar in these sprayed areas, and the fungicide shows up back in the colony (Fig. 3) at the levels that are approved for use in the field by the manufacturer. Bees also become contaminated while foraging and transport fungicides back to the colony on their bodies. Bees cover distances as far as seven miles to forage (Morse 1984), but will remain within four or five miles if there are enough flowering plants in the area. If bees are moved into orchards for pollination, they will frequently visit areas that have been sprayed, exposing them to different fungicides that can then be brought back to the colony. In fact, some of the highest levels of fungicides in bee bread and bee-associated products, like honey and wax, were found in bee colonies placed in certified organic fields (Yoder et al. 2013). Although there was no fungicide spraying occurring in the immediate vicinity, spraying occurred within the 3–5-mile range. Moving bee colonies to different orchards for pollination enhances this problem (Mullin et al. 2010), exposing the bees and their colonies to more and different kinds, mixtures and brands of fungicides, (and other pesticides)

Fig. 1 Sprayer truck applying fungicide in California almond orchards. *Photo* D. Sammataro

Fig. 2 Almonds and peaches in a California orchard. *Photo* D. Sammataro

Fig. 3 Skids of bee colonies for use as pollinators in California almonds. Skids allow for easy loading and unloading of colonies with a forklift. *Photo* D. Sammataro

with different concentrations, residual life, environmental persistence, and mode of action (Yoder et al. 2012a). Given the extensive use of fungicides in the USA and the broad area that bees can cover, the likelihood that bees will visit fungicide-sprayed areas is high. What effect are these fungicides having on the bee colony? And, what steps can be taken to maintain the strength and health of the bees?

Key to this chapter is the understanding that fungi, (which includes molds and yeasts) although not visible to the naked eye, are a vital part of the proper functioning of the bee colony. Essentially, the bee colony works as a large fermentation tank where the fermentation action of fungi is used to convert stored pollen into bee bread (Gilliam 1979, 1997; Gilliam et al. 1989; Gilliam and Vandenberg 1997; Vásquez and Olofsson 2009) and is also important for preserving the pollen for nutritional purposes (Anderson et al. 2014). The bee bread is then fed to developing bee larvae, and it is also an important protein source for growth of adult bees. Young nurse bees need to feed on bee bread to activate their food glands to feed bee larvae as well as the queen. Clearly, the amount and kinds of fungi that are in the colony are important for proper nutrition and for the normal functions of the bee colony. In addition, bee colony fungi play another critical role by providing immune protection as a natural resistance that protects the colony from infection. As such, having this battery of beneficial fungi puts the colony less at risk for bacterial, viral, fungal, and protozoal infections (Gilliam et al. 1988; Royce et al. 2015). Any factor or stress that disrupts the balance of these fungi has potentially harmful weakening effects, not only for an individual bee, but also on a much larger scale at the colony level. It is not just honey bee colonies affected, but all pollinating insects can be exposed and put at risk by fungicide applications (Bernauer et al. 2015).

Fungicides are involved with beekeeping practices because they kill, or decrease, the amount of fungi that are available in the environment, and consequently, the quantity of fungi inside the bee colony becomes less. This puts the colony at an elevated risk of disease because the natural defense shield provided by particular beneficial colony fungi could be weakened. Chalkbrood disease (caused by the fungal pathogen *Ascosphaera apis*; Fig. 4; Gilliam et al. 1988; Aronstein and Murray 2010) and pests such as wax moths (the greater wax moth *Galleria mellonella* and lesser wax moth *Achroia grisella*; Fig. 5) are two of the most common problems that can occur in colonies impacted by fungicides and are indicators that the colony has been weakened. Another problem is stonebrood disease (caused by the fungal pathogen *Aspergillus flavus*; Foley et al. 2014). It has been shown that fungicides weaken the bees and make them more vulnerable to nosema disease (caused by the pathogen *Nosema ceranae*) (Pettis et al. 2013). Fungi respond differently to different fungicides, and because there are many fungal strains, there is also considerable variation within the same species of fungus in relation to fungicide response. As such, two or more different strains of the same species of fungus could each respond differently to a specific fungicide (Bernert et al. 2012). Strain variation in fungi and variation in response to different fungicides have the end result of producing different mixtures of bee colony fungi, such that the ability of bees to fight disease also varies as well as colony nutrition.

This chapter aims to highlight the important health benefits of a good fungal balance inside the bee colony, showing visually what fungicides are doing to cause an imbalance, and focusing on subtle effects of fungicides on the colony fungi and how the bees are affected.

Fig. 4 Chalkbrood mummies on the ground from a colony weakened with fungicide. These mummies are the dead, hardened remains of bee larvae infected by the fungal pathogen *Ascosphaera apis. Photo* D. Sammataro

This chapter looks at:

1. locations of these important bee colony fungi;
2. types and quantity of fungi present in the bee colony; and
3. effects of fungicides on the colony environment that lead to disease.

Until alternative approaches to treating and preventing plant diseases with fungicides are devised and implemented on a large scale, the fungicide issue with bees will be an ongoing problem.

2 Origin of the Fungi in the Colony Environment

Fungi in bee bread and in honey bee colonies mostly originate from fungi in the bee's foraging habitat. Fungi get into the colony by way of spores. Numerous kinds and populations of fungi are found in the soil, where they function as agents of

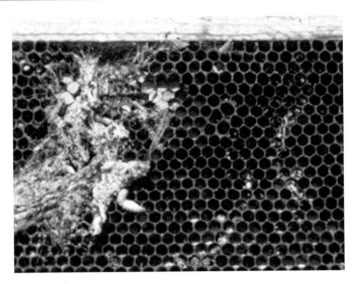

Fig. 5 Wax moth larvae on frames eating through the wax and causing damage to comb. This is seen in colonies not strong enough to keep the moths at bay. These moths are always around and will take advantage if the colony starts to decline from fungicide effects. *Photo* D. Sammataro

decay, and in or on plants (e.g., pollen and nectar). As fungi grow, they produce copious amounts of tiny spores that become airborne, spreading over plant surfaces. These spores are then carried back into the bee colony on collected pollen and nectar as well as on the bees themselves (Fig. 6). Once the pollen is packed into cells, the fungal spores can germinate and grow, fermenting the pollen and converting it into bee bread. Bee bread is an important protein source for bees that is used to activate the food glands in nurse bees to produce brood food as well as royal jelly. As the fungi grow within the column of packed pollen, the fungi interact and compete with each other through fungal-to-fungal interactions. Some fungi then become more dominant (while others are diminished), which ultimately gives rise to a mixture of fungi in different proportions (Jennings and Lysek 1999). The end result is bee bread that has a distinct fungal composition profile with a variety of fungal components in different amounts (Yoder et al. 2012b). In the fungicide-sprayed areas, the fungi are killed or fungal growth is reduced and there are fewer numbers of spores in the immediate environment. This means that there are fewer spores in the foraging habitat. The fungicide from spraying also coats the surface of pollen and becomes trapped in nectar, and fungicide residues also coat the bees when they visit flowers. When bees forage in fungicide-sprayed areas, the fungicide residues are also brought back into the colony on the bees themselves, on fungicide-contaminated pollen and nectar (Fig. 6; Kubik et al. 1999, 2000; Carlton and Jones 2007; Alarcón et al. 2009; Škerl et al. 2009; Yoder et al. 2013). Once

Fig. 6 Fungicides kill or stunt naturally occurring fungi in the bee's foraging habitat, so the number of fungal spores that regularly cover plant surfaces, pollen, and nectar that the bees collect and carry into the colony is reduced. These fungi are necessary for the operation and immune protection of the bee colony. Pollen and nectar are contaminated with fungicides, and the bees are coated with the chemicals. The fungicide is carried back to the colony and ultimately ends up in the bee's diet via bee bread from contaminated pollen. *Photos* J. Yoder, D. Sammataro, open access USDA-ARS Web site. Image construction (PSS Adobe Systems, San Jose, CA): A.J. Jajack and B.W. Nelson

packed into the honeycomb cells, the conversion from pollen to bee bread is hindered because of the diminished number of spores on the pollen. Additionally, fungicide residues on the pollen restrict fungal growth within the pollen that these fungi are trying to ferment.

A. Fungicide level: 17,000+ ppb (USDA ORGANIC)   **B. Fungicide level:** 0.00 ppb

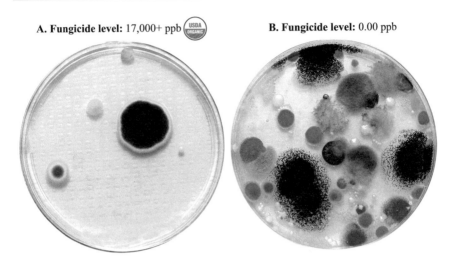

Fig. 7 Petri dishes showing cultures of fungi from bee bread. Bee bread samples were spread over the plate, and the fungi grew, producing round fungal colonies (each fungus colony grew from a single fungal spore). **a** Bee bread fungi from a certified organic orchard that showed high levels of fungicide. **b** Fungus colonies growing from bee bread taken from a bee colony that had no history of fungicide spraying and no detectable levels of fungicide. Conditions: potato dextrose agar, 0.5 mg bee bread sample, 2 days incubation, 30°C, darkness. *Photo* J. Yoder, from Yoder et al. (2013), J Tox Environ Health 76:587–600, reprinted with permission from Taylor & Francis Group, UK. Image construction (PSS Adobe Systems, San Jose, CA): A.J. Jajack and B.W. Nelson

3   Fungicide Effects in Bee Colonies

The fungus culture plates in Fig. 7 show evidence of fungicide on the beneficial bee colony fungi. Each fungal spore gives rise to a fungus colony when it is grown on agar in a Petri dish. In bee colonies exposed to fungicides, there are noticeably fewer fungi in bee bread even if that bee bread originated from a colony in a certified organic orchard which was in proximity to a fungicide-sprayed area (Fig. 7a; 17,328 ppb fungicide residues). Without fungicide exposure, there are numerous fungal colonies growing on the culture plate from a bee bread sample (Fig. 7b; 0.00 ppb fungicide residue detected in the bee bread sample by quantitative microanalytical chemical techniques). Even though this bee bread sample representing the fungicide exposure was taken from an apiary in an organic orchard (Fig. 7a), the amount of fungicide present was higher than samples taken from orchards that had been sprayed directly (Yoder et al. 2013). Seven different kinds of fungicides were detected in the organic sample, all at high concentrations.

4   Bee Colony Fungi and Fungicide Effects

The major categories of fungi found in bee bread and honey bee colonies in this study included high concentrations of the genus *Aspergillus* and *Penicillium*, with lower amounts of *Cladosporium* and *Rhizopus* (Fig. 8). There was also a mixture of

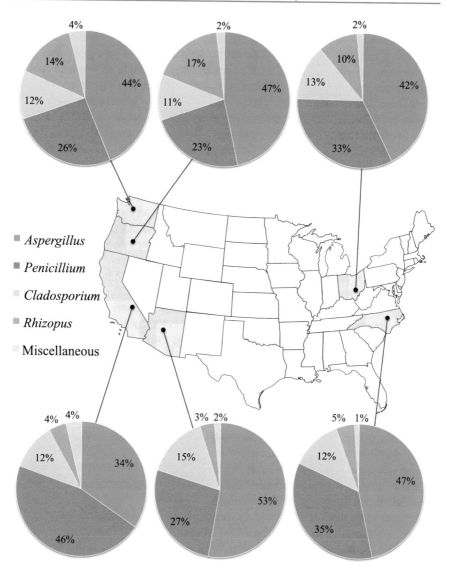

Fig. 8 Percentage of kinds of bee colony fungi from bee bread samples from different geographic regions. *Data* J. Yoder, 2015 unpublished, based on enumeration and fungal identification on standard media, potato dextrose agar, Melin-Norkrans agar, and Sabouraud agar (Fisher Scientific, Pittsburgh, PA), 30°C, darkness (Brown 2007; Barnett and Hunter 2003), using SPSS 14.0 for Windows (Microsoft Excel and Minitab; Chicago, IL) for statistical comparison. Scalable vector graphics (SVG) version of US state map is provided through Wikipedia's Creative Commons Attribution-Share Alike 3.0 Unported license. In accordance with the license terms, this figure is also being released under the Creative Commons Attribution-Share Alike 3.0 Unported license. Image construction (PSS Adobe Systems, San Jose, CA): A.J. Jajack

variable, miscellaneous components depending on what plants were present, climatic factors, and time of year of bloom. These variable fungal components included a mixture of *Absidia, Alternaria, Aureobasidium, Bipolaris, Fusarium, Geotrichum, Mucor, Mycelia sterilia* (sterile fungus, unable to produce spores), *Nigrospora, Paecilomyces, Scopulariopsis,* and *Trichoderma.* There are probably more fungi, but although our culturing techniques were general and equipped to handle most fungi, they did not recover everything (West et al. 2007). None of these fungi are unusual in bee colonies, and all are easily recognized fungi that are common in soil and plant settings (Jennings and Lysek 1999). Consistency in the kinds of bee colony fungi is one of the features that stand out, with the *Aspergillus/Penicillium* fraction dominating, regardless of geographic location. We attribute this consistency to bee colony conditions; the bees maintain the internal colony climate at a relatively stable 30–35 °C, 60–70% RH, and darkness (Cooper 1980; Chiesa et al. 1989) year-round. Since conditions inside the colony are relatively similar, the fungus profile is nearly the same wherever honey bee colonies are set-up (Fig. 8), with slight variations due to differences in foraging habitat.

Aspergillus, Penicillium, Cladosporium, and *Rhizopus* have been noted for quite some time to be important beneficial, disease-preventing fungi (Gilliam 1979; Gilliam et al. 1988, 1989; highlighted by Wood 1998; Aronstein and Murray 2010). High levels of these fungal components have been consistently found in bee colonies (Gilliam 1997; Gilliam and Vandenberg 1997; Osintseva and Chekryga 2008; Yoder et al. 2013; Foley et al. 2014). Some of these fungi produce antifungal compounds that kill or deter the growth of other fungi, while others can simply take over existing fungi by growing over them. The *Aspergillus/Penicillium* group has been associated with fighting off chalkbrood fungi (*A. apis*), which can kill bee brood. In particular, *Aspergillus niger*, one of the more dominant species in the *Aspergillus* group in bee bread, is reported in this regard (Royce et al. 2015). Bee bread that has high levels of *A. niger* is reported to protect against chalkbrood fungus (Royce et al. 2015). Other beneficial fungi in bee bread work together with *A. niger* to make it more efficient in fighting disease, so a combination of fungi is more beneficial than a single fungus alone.

Fungicide spraying has the effect of reducing all of these fungal components, and the *Aspergillus/Penicillium* group is hit particularly hard (Yoder et al. 2013). Fungicides commonly used tend to be particularly good at targeting this large group of fungi when tested against different species of *Aspergillus* and *Penicillium* individually (Yoder et al. 2012a). Bee colonies that show low levels of *Aspergillus/Penicillium* in combination with detectable levels of fungicides show chalkbrood symptoms (white 'chalk'-like brood in the frames of honeycomb as well as on the ground in front of the beehive; Fig. 4). As previously mentioned, numerous parasitic diseases are noticed in colonies that are exposed directly or indirectly to fungicide spraying. So, these beneficial bee colony fungi can be tied to the immune protection against disease for the entire colony. Lowering the amount of these fungi can leave the colony open to attack by other pathogens and parasites. Fungicides have been suggested to be one of the factors in colony collapse disorder (CCD). We feel that the linkage between fungicides and CCD is that colonies with lower beneficial fungi (*Aspergillus/Penicillium*), killed by fungicides, make the colony immunocompromised.

5 Observations of Bee Colonies in Fungicide-Sprayed Areas

Ecologically, just because a honey bee colony has a low number of fungal components does not mean that it is not active or functional. Bees have a remarkable way to adapt to their environment, and the amount and kinds of fungi that are present in the colony at a particular location is what works best for that bee colony. The composition and levels of beneficial fungi (different species, or even strains of *Aspergillus/Penicillium*) for one colony in one habitat may be different from colonies in a different habitat. The key feature we found is not to disrupt the fungal balance inside the colony. Fungicides appear to disrupt the balance, and that compromises colonies in that particular environment. This is especially true for migratory beekeeping operations, which move thousands of colonies in and out of orchards and crops as they bloom. Bees are exposed not only to different fungi in the environment, but also to different pesticides (including fungicides) as they are moved from crop to crop. Addition of beneficial probiotic fungi to the colony is not advised. Healthy bees maintain the level in the colony that works best for them and to be able to function effectively in that environment. One of the best solutions is to keep bees healthy which allow the fungi and bees to take care of each other.

In all of the colonies that we sampled, most of them were active, thriving colonies, including ones that were in heavily fungicide-sprayed areas or contained a lot of fungicide inside the colony. Bees appeared to develop normally, at least short-term, during a period of 3–6 months. Even with the reduced load of bee bread fungi (Fig. 7a), this low amount of fungi is apparently still capable of converting pollen into bee bread that is viable, because the bee colonies in our study were active. It is possible that the fungi are not the sole group involved in fermentation, or that they are the first group in a succession pattern. In the fungicide-exposed colonies, however, there was increased incidence of disease, mainly chalkbrood. So, at the concentrations and frequency of fungicide applications that are recommended for use in the field by manufacturer, the reduction in beneficial colony fungi does not appear to negatively impact the colonies, at least in the short-term. What the long-term effects are and whether fungicides are causing nutritional stress to the colony (Naug 2009) are not known. In addition, the effect of fungicides on other pollinating insects is just now being studied (Bernauer et al. 2015).

6 Protecting Bees from Fungicide (and Other) Sprays

The ultimate responsibility to protect honey bee colonies rests with the beekeeper. The timing of fungicide spraying in the evening hours, when bees are not foraging, does not seem to work well as far as preventing fungicide contamination to bee colonies (Mullin et al. 2010). This is because of the residues that fungicides leave behind, residual life, the persistence in the environment, and the indirect effects from nearby sprayed fields. Fungicide contamination can take place within a matter of weeks and put a bee colony in trouble. What we have presented in this chapter only concerns the short-term effects on the bee colony. No long-term studies have been done, and there are still many unresolved questions. Fungicides have been labeled safe for bees, but our observations and information by beekeepers have shown that these compounds can cause problems. Bee larvae, in particular, have been reported to be killed by fungicides (Alarcón et al. 2009; Mussen 2008).

In summary, we found:

1. Even at the approved concentrations of fungicides and manufacturer's directions that are being applied in the field, there is a sufficient amount inside honey bee colonies that can decrease the amount of fungi present in bee bread.
2. Bee colonies placed in certified organic fields but surrounded by fields that are sprayed will collect fungicide-contaminated pollen, because the bees will visit nearby sprayed fields. Bees are not confined to a particular field if there are other (sprayed) orchards in bloom.
3. Fungicide spraying in the USA is extensive. In the hundreds of bee bread, honey, and bee-associated samples, we have examined from across the country, ranging from large-scale bee operations to small scale, farm and homeowner beekeeping with only a few colonies, we found only one yard (Arizona) that had no detectable amount of fungicide. This was our control apiary. This control apiary was the one at the USDA in Tucson, AZ. This apiary became our control apiary when bee bread and bee colony products showed no fungicide residues (0 ppb), when samples from these colonies were tested.
4. Chalkbrood, stonebrood, and nosema diseases, and parasites can be signs that the colony is suffering from effects of fungicide contamination.
5. Fungicides, with the amounts that we found present in the field, are having an appreciable impact on lowering the bee colony's ability to fight off infection, which sets up the colony to be taken over by bacterial and viral diseases, and parasites.

It is extremely important to note that these negative effects of fungicides on disease-prevention can intensify and synergize with other pesticides and miticides (Johnson et al. 2010, 2013; Bernauer et al. 2015). The concern with acaricides is pressing, especially, because of their more widespread use to combat the Varroa mite (*Varroa destructor*). So, acaricides and pesticides can exacerbate the stress imposed by fungicides on compromising the bee colony in terms of disease defense.

Here are some steps that can be taken to alleviate bee colony exposure to fungicides and other agrochemicals.

Prior to setting up a bee colony or apiary:

1. Attend local beekeeper or grower association meetings to talk and discuss the potential issues with other area beekeepers and orchardists. It is important to

understand the events that are occurring in the bee's foraging habitat. Encourage local clubs to contact sprayers to make them aware of the potential problems.

2. Simply drive around the proposed site and note the crops. Map out the best site to set up an apiary with minimal fungicide exposure within a 3–5-mile radius of the intended apiary and ask about what crops are grown, the production methods, past pest or disease problems, planting, blooming, and harvest dates. The best way to do this is to become active in local beekeeping organizations. Some local beekeeping guilds have programs where they put the GPS locations on the web and local sprayers check it in relation to their spraying. Sometimes local sprayers even change their schedule so the wind blows sprays away from hive areas. Become knowledgeable with spraying schedules and what fungicides (and other pesticides) are used for that orchard, crop, or location. Check plat, county, or air photomaps or online maps, even soil type maps, to assess apiary location in relation to the areas that may be sprayed (parks, orchards, tree farms, and residences). Even for migratory beekeepers, it will help them to be aware of the pesticides used in the area where their bees are located.

3. Be aware of the pesticide danger potentials: spray drift from nearby treated areas, frequency of sprays during the season, cyclic or unexpected outbreaks of insect pests or plant pathogens that would include the use of insecticides or fungicides, the need for sprays during the blooming period, and application methods (air or ground, low volume, ultra-low volume, standard, or electrostatic equipment). Often, spray applications are mixed, such as fungicides and insecticides; it is critical to know what chemicals are being used. Conduct an Internet search on particular pesticides for valuable information about additional concerns and issues that might have not been considered.

4. Find out and know the crop pests and plant pathogens (or insect pests for insecticides) that are in the area. Know when other chemicals will be applied (e.g., after a rain). Ask about the types of fungicides that are used locally, their common names and formulations.

Before Fungicide Spraying

Move colonies at least 2–3 miles from their previous location and target area. If colonies can not be moved, make sure the applicators (local or contracted) know the locations of the apiary locations (supply maps). Name, address, and phone numbers of the beekeeper should be conspicuously posted on colonies or near the apiaries. Paint hive tops with a light color for easy aerial identification.

When Fungicide Spraying is About to Happen

1. Reduce hive entrances to restrict number of bees flying out.
2. Gorge hives with sugar syrup, by pouring it directly on top of the frames (bees will stop foraging to help clean it up); pour in about a quart of syrup twice a day for one day prior to a spray, and once a day for 2–3 days following a spray.
3. Close the hive entrances with eight-mesh hardware cloth or screen to confine bees, and place a screened cover on top, covered with a wet cloth or wet

Fig. 9 Trapping pollen to reduce the impact of fungicide-contaminated pollen that gets into the colony. *Photo* D. Sammataro

burlap. To keep the burlap wet, especially if the weather is hot, use a sprinkler or watering can, for *at least* 24 h. This is a dangerous step to take, because even with wet burlap, hives could overheat and die very quickly. If weather is cool, only the eight-mesh hardware cloth screened entrance is needed.

4. Activate pollen traps to collect contaminated pollen (destroy this pollen afterward) (Fig. 9).
5. Feed colonies with syrup, water, and clean pollen patties (Fig. 10) during *and* after the spray period.

After Fungicide Spraying

1. Look for signs that the colony is in decline, such as low populations of bees and brood, missing queen, and dying bees at the colony entrance. To save colonies, replace contaminated stores, combs, and equipment with new equipment, clean combs, and foundation. Feed sugar syrup, pollen, or pollen substitutes to maintain colony populations and to stimulate brood rearing.
2. Monitor bee colony fungi to look for drastic changes in fungus levels. To collect bee bread, use a coring technique by inserting a disposable, sterile, polyethylene

Fig. 10 Feeding clean pollen or pollen substitute to help weakened colonies recover from fungicide effects. *Photo* D. Sammataro

pipette tip (Fisher Scientific, Pittsburgh, PA) or clean, prepackaged soda straws, directly into the center of bee bread honeycomb cells. Collect one core sample per cell, and collect from multiple (5 to 10) cells and several frames per colony. The bee bread (or stored pollen) core sample remains inside the tip or straw. Labeling is important, so indicate site, date, colony number (so the individual colony can be tracked), any notes about the colony, treatment histories, and position in the frame from where each core sample was collected. The tips or straws containing the core sample can then be placed into plastic bags or wrapped in foil and sent off for analysis (university, college, extension offices, or natural resources departments). Some USDA bee laboratories will also determine pesticide levels. Check with local laboratories before sending samples, because not all laboratories will do this, some will charge fees, and the laboratories need to have some knowledge beforehand on how to process and interpret the data. Fungi can be determined by culturing and fungal colony enumeration techniques in any standard microbiology laboratory (Royce et al. 2015). Overnight or fastest shipping is preferred. It is important *not* to put the samples in the freezer, as freezing can kill some fungi that would change its fungal profile once it is analyzed. The same core samples can also be used for fungicide and pesticide residue analysis.

3. Add new frames of bees to colonies to increase the genetic diversity of the colony that has been shown to improve colony health through a greater diversity of beneficial fungi as well as other microorganisms (Mattila et al. 2012).

Stay Current on the Latest Developments as Spraying Policies Change
Beekeepers should strive to cooperate with neighboring growers for their mutual benefit. Growers need bees to pollinate and beekeepers need growers. There are several places to get more information on the location of orchards, chemicals currently used, and possible effects on bees. These include:

1. State and county extension offices (extension entomologists, agronomists, horticulturists), and their publications on crops, pests, and diseases in the area. Check Web sites as well.
2. State apiary inspectors.
3. Regional, state, and local beekeeping organizations.
4. Libraries or city/state agencies, agriculture departments. Check their Internet sites for maps and other references.

Acknowledgements Special thanks to Dr. Michael A. Senich (Midland, TX) for funding to BWN (Wittenberg University, Springfield, OH).

References

Alarcón R, DeGrandi-Hoffman G, Wardell G (2009) Fungicides can reduce, hinder pollination potential of honey bees. Western Farm Press 31:17–21

Anderson KE, Carroll MJ, Sheehan T, Mott BM, Maes P, Corby-Harris V (2014) Hive-stored pollen of honey bees: many lines of evidence are consistent with pollen preservation, not nutrient conversion. Mol Ecol 23:5904–5917

Aronstein KA, Murray KD (2010) Chalkbrood in honey bees. J Invert Pathol 103:S20–S29

Barnett HL, Hunter BB (2003) Illustrated genera of imperfect fungi, 4th edn. APS Press, St. Paul

Bernauer OM, Gaines-Day HR, Steffan SA (2015) Colonies of bumble bees (*Bombus impatiens*) produce fewer workers, less bee biomass, and have smaller mother queens following fungicide exposure. Insects 6:478–488

Bernert AC, Sagili RR, Johnson KB (2012) Evaluating pesticide sensitivity of the honey bee (*Apis mellifera*) microbiome. ESA Annual Meeting Online, Knoxville

Brown AE (2007) Benson's microbiological applications: laboratory manual in general microbiology. McGraw-Hill, New York

Carlton AJA, Jones A (2007) Determination of imidazole and triazole fungicide residues in honeybees using gas chromatography-mass spectrometry. J Chromatogr A 1141:117–122

Chiesa F, Milani N, D'Agaro M (1989) Observations of the reproductive behavior of *Varroa jacobsoni* Oud.: techniques and preliminary results. In: Cavalloro R (ed) Proceeding of the meeting of the EC-Experts' Group, Udine 1988, pp 213–222

Cooper B (1980) Fluctuating broodnest temperature rhythm. Br Isles Bee Breeders News 18:12–16

Foley K, Fazio G, Jensen AB, Hughes WOH (2014) The distribution of *Aspergillus* spp. opportunistic parasites in hives and their pathogenicity to honey bees. Vet Microbiol 169:203–210

Gilliam M (1979) Microbiology of pollen and bee bread: the yeasts. Apidologie 10:43–53

Gilliam M (1997) Identification and roles of non-pathogenic microflora associated with honey bees. FEMS Microbiol Lett 155:1–10

Gilliam M, Vandenberg JD (1997) Fungi. In: Morse RA, Flottum PK (eds) Honey bee pests, predators and diseases, 79–112

Gilliam M, Taber S III, Lorenz B et al (1988) Factors affecting development of chalkbrood disease in colonies of honey bees, *Apis mellifera*, fed pollen contaminated with *Ascosphaera apis*. J Invert Pathol 52:314–325

Gilliam M, Prest DB, Lorenz BJ (1989) Microbiology of pollen and bee bread: taxonomy and enzymology of molds. Apidologie 20:53–68

Jennings DH, Lysek G (1999) Fungal biology: understanding the fungal lifestyle. Springer-Verlag, New York

Johnson RM, Ellis MD, Mullin CA, Frazier M (2010) Pesticides and honey bee toxicity—USA. Apidologie 41:312–331

Johnson RM, Dahlgren L, Siegfried BD et al (2013) Acaricide, fungicide and drug interaction in honey bees (*Apis mellifera*). PLoS ONE doi:10.1371/journal.pone.0054092

Kubik M, Nowacki J, Pidek A, Warakomska Z, Michalczuk L, Goszczyñski W (1999) Pesticide residues in bee products collected from cherry trees protected during blooming period with contact and systemic fungicides. Apidologie 30:521–532

Kubik M, Nowacki J, Pidek A, Warakomska Z, Michalczuk L, Goszczyñski W, Dwuznik B (2000) Residues of captan (contact) and difenoconazole (systemic) fungicides in bee products from an apple orchard. Apidologie 31:531–541

Mattila HR, Rios D, Walker-Sperling VE, Roeselers G, Newton ILG (2012) Characterization of the active microbiotas associated with honey bees reveals healthier and broader communities when colonies are genetically diverse. PLoS ONE 7:e32962

Morse R (1984) Research review: How far will bees fly? Gleanings in Bee Culture September: 474

Mullin CA, Frazier M, Frazier JL, Ashcraft S, Simonds R, vanEnglesdorp D, Pettis JS (2010) High levels of miticides and agrochemicals in North American apiaries: implications for honey bee health. PLoS ONE 5:e9754

Mussen E (2008) Fungicides toxic to bees? Apiculture Newsletter Nov/Dec 2008

Naug D (2009) Nutritional stress due to habitat loss may explain recent honeybee colony collapses. Biol Conserv 142:2369–2372

Osintseva LA, Chekryga GP (2008) Fungi of melliferous bees pollenload. Mikol Fitopatol 42:464–469

Pettis JS, Lichtenberg EM, Andree M et al (2013) Crop pollination exposes honey bees to pesticides which alters their susceptibility to the gut pathogen *Nosema ceranae*. PLoS ONE 8 (7):1–9. doi:10.1371/journalpone.0070182

Royce L, Yoder J, Nelson B et al (2015) Tree hive colonies: increased quantity of beneficial fungi in bee bread from the trees and its antifungal properties against chalkbrood. Bee Culture March: 59–63

Škerl MIS, Velikonja Bolta S, Baša Česnik H, Gregorc A (2009) Residues of pesticides in honeybee (*Apis mellifera carnica*) bee bread and in pollen loads from treated apple orchards. Bull Environ Contam Toxicol 83:374–377

Vásquez A, Olofsson T (2009) The lactic acid bacteria involved in the production of bee pollen and bee bread. J Apic Res 48:189–195

West SA, Diggle SP, Buckling A et al (2007) The social lives of microbes. Ann Rev Ecol Evol Syst 38:53–57

Wood M (1998) Microbes help bees battle chalkbrood. Agric Res 46:16–17

Yoder JA, Hedges BZ, Heydinger DJ et al (2012a) Differences among fungicides targeting beneficial fungi associated with honey bee colony. In: Sammataro D, Yoder JA (eds) Honey bee colony health: challenges and sustainable solutions, 181–192

Yoder JA, Heydinger DJ, Hedges BZ et al (2012b) Fungicides reduce symbiotic fungi in bee bread and the beneficial fungi in colonies. In: Sammataro D, Yoder JA (eds) Honey bee colony health: challenges and sustainable solutions, 193–214

Yoder JA, Jajack AJ, Rosselot AE, Smith TJ, Yerke MC, Sammataro D (2013) Fungicide contamination reduces beneficial fungi in bee bread based on an area-wide field study in honey bee, *Apis mellifera*, colonies. J Tox Environ Health A 76:587–600

Author Biographies

Jay A. Yoder is a Professor of Biology at Wittenberg University in Springfield, Ohio. He teaches courses in microbiology and immunology and general biology. Topics of his research include the physiology of ticks and mites and stream invertebrates in addition to honey bees, with an emphasis on water balance, chemical ecology, and fungal relationships. Most of his work has been done with the help of undergraduates (Wittenberg does not have a graduate school), amounting to over 170 peer-reviewed publications in scientific journals, and book chapters, with undergraduate students as co-authors. Jay holds a B.A. from the University of Evansville (biology, chemistry, and French), a Ph.D. from The Ohio State University (Entomology; laboratory of Dr. David Denlinger) and conducted post-doctoral work at Harvard University (laboratory of Dr. Andrew Spielman).

Diana Sammataro is co-author of the Beekeeper's Handbook, began keeping bees in 1972 in Litchfield, Connecticut, setting up a package colony in her maternal grandfather's old beehive equipment. From then on, she decided that her B.S. in landscape architecture (University of Michigan, Ann Arbor), would not be a career, but that honey bees would. After a year of independent studies on floral pollination (Michigan State University Bee Lab, East Lansing), she earned an M.S. in Urban Forestry (University of Michigan, Ann Arbor). In 1978, she joined the Peace Corps and taught beekeeping in the Philippines for 3 years. On returning, she worked at the USDA Bee Lab in Madison, Wisconsin, under Dr. Eric Erickson, studying the effects of plant breeding and flower attraction of bees in sunflower lines. When the laboratory closed, she eventually went to work at the A.I. Root Company as Bee Supply Sales Manager in Medina, Ohio. In 1991, she was accepted at the Rothenbuhler Honey Bee Lab at The Ohio State University, Columbus, to study for a Ph.D. under Drs. Brian Smith and Glen Needham. In 1995, she worked as a post-doctoral assistant at the Ohio State University Agricultural Research Center in Wooster, Ohio, with Dr. James Tew and in 1998 at the Penn State University Bee Lab. Early in 2002, she was invited to join the USDA-ARS Carl Hayden Honey Bee Research Center in Tucson, Arizona. There she worked as a Research Entomologist until 2014, looking at bee nutrition problems, how they influence Varroa mites, and current pollination problems.

Using Honey Bee Cell Lines to Improve Honey Bee Health

Michael Goblirsch

Abstract

Through their pollination services, honey bees serve an essential role in sustaining the nutrition, health, and shelter of humans and other animals. Because of the demand for their pollination services, beekeepers strive to keep honey bees healthy and productive. Unfortunately, many factors commonly found throughout the USA can detract from honey bee health. For example, infections with pathogens such as viruses and *Nosema* spp. pose a significant threat to honey bee survival. Viruses and *Nosema* spp. must invade cells of their host to reproduce and complete their development. To better understand the interaction between intracellular pathogens and honey bee cells, tools are needed that allow examination of infection at fine resolution and under controlled, aseptic conditions. One tool that has been underutilized in honey bee biology is cell culture; therefore, the focus of this chapter is to discuss the technique and use of honey bee cells in culture. Infection of honey bee cells in culture with the fungal pathogen, *Nosema ceranae*, is used as an example to demonstrate how cell culture can be a powerful tool to explore the process of infection and the negative impact pathogens may have on honey bee biology and health.

M. Goblirsch (✉)
Department of Entomology, University of Minnesota, 219 Hodson Hall,
1980 Folwell Ave, St. Paul, MN 55109, USA
e-mail: goblirmj@umn.edu

© Springer International Publishing AG 2017
R.H. Vreeland and D. Sammataro (eds.), *Beekeeping – From Science to Practice*,
DOI 10.1007/978-3-319-60637-8_6

1 From Honey Bee Decline Comes Opportunity

The emergence of Colony Collapse Disorder (CCD) created a wave of opportunities for beekeepers, researchers, and policymakers to come together to identify the roots of honey bee decline, to develop innovative and practical measures to reverse the negative trend in bee health and productivity, and to support training of the next generation of apiculturists (i.e., beekeepers and researchers) (Pollinator Health Task Force 2015). One area that has seen tremendous growth in response to CCD is our understanding of the *micro-managers* of honey bee biology. These micro-managers include the DNA, RNA, proteins, and other factors housed within the cells of a bee. Our everyday awareness of these micro-managers may not impact how we perceive a colony of bees, but in essence, they run the show by carrying out cellular processes such as metabolism, detoxification, immunity, and tissue repair to keep bees functioning properly. The synthesis, performance, and degradation of a bee's micro-managers are partly shaped by the bee's day-to-day interactions with its environment. These interactions take place at many levels and can include the exchanges a bee has with its nestmates in the hive, the flowers it visits on a foraging trip, the nectar or pollen it consumes, as well as its exposure to pathogenic microbes and pesticides.

Advances in molecular biology have led to the development of rapid, accurate, and reproducible techniques for quantifying the micro-managers of a honey bee's life. It is now possible to collect bees from a colony, dissect them down to their most basic elements, and compare, for example, differences in the levels of RNA or proteins between bees that are healthy and ones that are diseased. This information can then be used to develop strategies or therapies for improving the health of a colony inflicted with disease, such as supplementing its nutrition, providing treatments, or other approaches. However, to gain a deeper understanding of the effects of disease on bees, it would be advantageous to observe these effects at the level of where they occur, the cell. Unfortunately, our use of cell culture for the purpose of understanding and manipulating honey bee health lags behind other insects of economic or medical importance. The culture of cells from insects that cause human or livestock diseases or pests that damage crops have been refined over decades and are well established. Conversely, very few research programs use honey bee cells or tissues in culture to explore current problems that affect the health of this important beneficial insect.

The objective of this chapter is to convey the importance of cell culture in honey bee research. To achieve this objective, a brief history and description of the technique of cell culture is given, using previous work done with honey bee cells. I then discuss the process involved in the development of a line of continuously dividing cells from honey bee embryonic tissues (Goblirsch et al. 2013). How bee cells can be used to study pathogens that cause bee disease is discussed, using the fungal pathogen, *Nosema ceranae*, as an example. *Nosema ceranae* is only part of the story of how honey bee cell culture can be used to explore issues in bee biology and disease. Honey bee cell culture is a powerful tool that permits examination, at fine resolution and under controlled, aseptic conditions, those factors that cause a bee to succumb to pathogens or other stressors. Once the effects of these factors, whether they are pathogens, pesticides, or nutritional deficiencies, are identified and described, we can apply an integrated approach from disciplines such as micro-, cellular, and molecular biology, as well as beekeeping practices to help restore the health and productivity of our honey bees.

2 A Brief History of Honey Bee Cell Culture

Cell culture has an extensive history in insect biology and disease. Cell culture has advanced our understanding of how the different cell types of an insect (such as the muscle, nerve, blood, or cuticular cells) interact with each other and contribute to the formation and function of complex systems, such as tissues. It has been more than a century since some of the earliest research on insect cell culture. Early attempts at culturing insect cells involved extracting tissues from the whole organism and placing these cells on glass slides in a salt or similar solution. In most cases, the cells were short-lived (e.g., hours or days), but these cultures provided basic information on processes such as cell division (Goldschmidt 1915). From this early research began a progression toward the development of more suitable media to replace saline to fulfill the nutritional requirements of insect cells in culture. Identification and levels of sugars, amino acids, vitamins, minerals, and growth factors, such as hormones, that mimic the conditions of those from where the cells were extracted lead to the establishment of cultures with increased cell viability and proliferation.

A breakthrough occurred in the 1960s when Grace (1962) isolated the first line of continuously dividing insect cells from the ovarian tissues of the Emperor Gum Moth, *Antheraea eucalypti*. Since this achievement, over 500 insect cell lines have been developed, mainly from tissues of different species of flies (Diptera), leafhoppers (Hemiptera), or moths/butterflies (Lepidoptera), but very few from wasps, ants, and bees (Hymenoptera) (Lynn 2003). The motivation for the development of cell lines from insects, such as flies or moths, was due to several of these insects being significant vectors of disease in humans, livestock, and plants. For example, *Anopheles* mosquitoes transmit *Plasmodium falciparum*, the causative agent of the most severe form of malaria. Cell lines established from *Anopheles gambiae* mosquitoes have been used to develop models of malaria transmission. From this model cell culture system, researchers have been able to elucidate mechanisms of pathogen–host interactions, such as the suppression of the immune system of the mosquito caused by the protozoan that allows for maintenance of the infection in the insect vector (Walker et al. 2014).

Insect cell lines have also lead to novel approaches for pest control, notably the mass production of insect viruses in cultured cells for use as *biopesticides* against crop pests (Arif and Pavlik 2013). The utility of insect cell lines has changed dramatically after it was discovered that a specific group of insect-infecting viruses, the baculoviruses, could deliver genes from different organisms into insect cells and drive the production of novel proteins (Smith et al. 1985). Many insect cell lines are now grown in large batches for the manufacture of proteins that have an impact on human and animal health, most notably vaccines (Airenne et al. 2013).

So, why is a honey bee cell line needed? The functional capacity of existing insect cell lines is rapidly expanding. Fueled by advances in biotechnology and insect pathology comes the need to develop additional continuous insect cell lines. Ironically, the celebrity-like status of the honey bee has not propelled it to the forefront as a cell culture model. Until recently, no cell lines were reported from the honey bee (Kitagishi et al. 2011; Goblirsch et al. 2013; Ju and Ghil 2015). Furthermore, studies demonstrating the use of honey bee tissues to initiate cell cultures are surprisingly few. Primary, or short-lived, cell cultures have been established using honey bee eggs (Giauffret et al. 1967; Bergem et al. 2006; Chan et al. 2010) and larval and pupal tissues (Stanley 1968; Giauffret 1971; Beisser et al. 1990; Gascuel et al. 1994; Gisselmann et al. 2003), but the longevity of these cultures was less than three months. Primary cultures from neural tissues are an exception to an otherwise sparse literature based on bee cell culture. Neurons extracted from brain structures responsible for processing sensory information have proved valuable for increasing our understanding of the development, form, and function of the honey bee olfactory system (Kreissl and Bicker 1992; Bicker and Kreissl 1994; Schäfer et al. 1994; Goldberg et al. 1999; Kloppenburg et al. 1999; Grünewald 2003; Malun et al. 2003; Barbara et al. 2008).

3 Cell Culture Defined

Cell culture, sometimes referred to as tissue culture, may not be familiar to many, so an explanation of this widely used laboratory technique is given here using honey bees as a model. Honey bee cell culture involves the extraction of a fragment of tissue or the purification/separation of cells from the blood of the bee under sterile conditions. Depending on the type of questions being asked, certain tissues are better than others for establishing a bee culture system. For example, muscle cells of the bee could be removed from the thorax and used to examine factors that influence their growth, repair, or, from an applied perspective, the metabolic demands of honey bee flight. However, muscle cells are *terminally differentiated*, meaning they have reached their final form and are committed to the function of movement or maintaining posture. As a result, these cells can no longer divide; therefore, a culture composed of this cell type has a finite life span because its cells have lost the ability to replenish the cell population. If the goal is to establish a culture that contains cells with a high potential for replication, then targeting tissues or life stages where there is a high rate of turnover and rejuvenation in the cell population would be more suitable. Examples of bee tissues or life stages that contain cells with a high rate of cell division are the ovaries of the queen, the midgut cells of workers, wing disks of larvae, or honey bee eggs.

Once a tissue is identified that will satisfy the needs of the experimental design, a fragment of the tissue or isolate of the blood is extracted from the bee and introduced into an artificial environment. The artificial environment is typically a sealable plastic flask that can be placed in an incubator regulated for temperature, humidity, and atmospheric gases (e.g., concentration of carbon dioxide). It is crucial that the environment the bee cells are introduced into is free of contamination from microbes (such as bacteria, fungi, protozoa, and viruses) and cells from organisms other than the species from where the fragment or isolate originated.

4 Honey Bee Cells Need a Medium to Grow

Honey bees are composed of different cell types (e.g., gland, muscle, nerve, or blood cells) that are organized into tissues and, at a higher level, systems. Within an intact bee, cells that make up different tissues are in contact, interact, and communicate with each other and with cells from other tissues and systems. Removing and placing cells in a plastic flask puts them in an unfamiliar, if not hostile, environment. To help bee cells adapt and survive under these conditions, a growth medium is added to the flask. The growth medium is a nutrient-rich liquid tailored to support fundamental metabolic processes and approximate the physiological conditions from which the cells were extracted. The composition of the medium varies depending on the species of origin and the type of tissues or cells brought into culture. For example, the medium used to culture bee cells contains amino acids for protein synthesis, lipids, and sterols for plasma membrane formation, carbohydrates as a source of energy, vitamins and growth factors to support viability and proliferation, and minerals for ion exchange and catalysts in chemical reactions. The amounts of these constituents that are added to the medium have been determined from chemical analyses of tissues and blood collected from insects of different life stages or exposed to variable conditions such as different incubation temperatures.

Honey bee cells in culture are bathed in the growth medium and, from this, acquire nutrients and other factors needed to remain viable and *potentially* proliferate via cell division. Figure 1 depicts a honey bee cell undergoing division to produce two daughter cells. The blue arrow in panel A denotes the phase in the cell cycle called metaphase. Metaphase is the stage of cell division where the chromosomes of the original cell have doubled to form pairs that are aligned along the equator of the cell, prior to being separated into the two daughter cells. The yellow arrow in panels B and C depicts the constriction of the plasma membrane around the cytoplasm to form daughter cells, each with its complement of chromosomes. It takes approximately 45 min for a honey bee cell at room temperature to complete one round of cell division.

It is important to emphasize the term *potentially* with regard to division for bee cells in a primary culture because there is no guarantee that these cells will divide. The inability to undergo cell division may be due to a number of reasons,

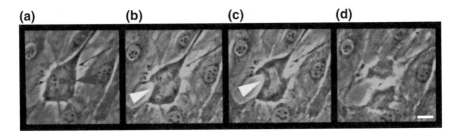

Fig. 1 An AmE-711 honey bee cell undergoing division. **a** A parent honey bee cell in late metaphase/early anaphase of the cell cycle where the chromosomes have duplicated to form pairs of sister chromatids. The pairs of chromatids are lined up along the *equator* of the cell, marked with a *blue arrowhead*. The pairs of chromatids will eventually split, and the sister chromatids will migrate toward opposing poles of the cell. **b** The *yellow arrowhead* points to the cleavage furrow, which shows the stage in cell division where the cell membrane begins to constrict or pinch the parent cell into two daughter cells. At this stage in cell division, the membranes are beginning to reform the nucleus around the sister chromatids that migrated previously to each daughter cell. The cleavage furrow is more pronounced in **c**. **d** The original parent cell has completed division to form two daughter cells, which are marked with *gray arrows*. The process of cell division from start to completion takes approximately 45 min at room temperature. Scale bar = 10 μm

but factors absent from the medium or characteristics inherent to the cells them-selves likely contribute to this failure. For example, when cells are present in the body of the bee they respond to hormones and growth factors in circulation. Some hormones and growth factors serve as signals that prompt cells to divide. Our knowledge of the identity and function of many of these factors in honey bees is incomplete and may partly explain why there has not been more progress in establishing continuous bee cell lines.

5 Simple Cell Cultures as an Alternative to Complex Colonies

Honey bees are susceptible to infection from several different pathogenic microorganisms. Some of these microbes are obligate intracellular pathogens, meaning they must invade a host cell to reproduce and complete their development. Examples of these pathogens that pose problems for bees are *Nosema* spp. and viruses. Much of what we know about the host–pathogen dynamics of a recent pathogen found to infect honey bees, *Nosema ceranae*, has been ascertained from time course studies where bees, typically workers, are fed spores and then placed in cages or colonies to observe the outcomes of infection. A drawback to caged bee or colony studies is that these experimental units can be *messy*. Factors difficult to control, especially for colonies (e.g., fluctuation of weather and temperature con-ditions, nutritional status of bees, presence/absence, and levels of pheromones that regulate worker development), can mask the true effects of the intended treatment variable. Paradoxically, studies conducted in cages and colonies offer a real-world presentation of how detractors to honey bee health, such as disease caused by *N. ceranae*, can alter the response of bees at the organismal and social levels. As beekeepers and researchers, we are constantly aware of the unpleasant reality that our bees exist in a world wrought with contamination from pesticide residues and pathogens, exploitation by mite pests, and nutritional deficiencies due to foraging in impoverished landscapes. To understand the impact a single factor has on honey bee health, for example, the interaction of *N. ceranae* with its host, it would be ideal to eliminate as many of the confounding effects as possible. This is where cell lines derived from honey bee tissues can come to the rescue. Honey bee cell lines are *clean* systems composed of homogenous units maintained in a regulated environment. Moreover, bee cell lines can be utilized to observe how a single factor affects the biology of the bee at a fine scale of resolution. Importantly, honey bee cell lines also allow observation of emerging pathogens in a system that is contained. This level of control is not possible with colonies where bees leave the hive freely and could potentially spread disease to other colonies or to native bees.

6 Establishing Primary Cultures from Honey Bee Tissues

To supplement the excellent data available from studies using colonies and cages, we developed a culture system using bee cells that would allow collection of detailed, controlled data on infection with pathogens inside its host cells. Techniques and formulations of growth medium used to initiate primary cultures from other insects offer a good baseline to attempt the culture of bee cells. The first step in the process of initiating primary cultures is to identify tissues that would be relatively easy to bring into culture and have the greatest potential to survive for an extended period. The literature on honey bee development and beekeeping management points to the egg as an excellent source of donor tissue for several reasons. First, it has been shown previously that actively dividing honey bee cells can be maintained for more than three months using eggs (Bergem et al. 2006). Second, given that a healthy queen lays hundreds of eggs per day in the summer, it is relatively easy to take a frame from a colony in the field and bring it to the laboratory to collect a large number of eggs. Third, the smooth, ovoid form of the egg requires less effort to sterilize its surface compared to the convoluted tissues of the larva, pupa, or adult. Finally, and perhaps most importantly, a honey bee egg just prior to hatching contains cells that have fully formed membranes and are progressively dividing (DuPraw 1967).

By caging a queen on a frame of empty comb into the center of a brood box for 24 h, it is possible to obtain a large number of eggs that are synchronized in their development (Fig. 2). To increase the chances that a queen will lay eggs on the frame where she is caged, the frame must first be inserted into the colony to allow the workers to clean the comb cells in preparation for her to lay. After 24 h, the queen can be caged to one side of the frame using hardware cloth. The partitions of

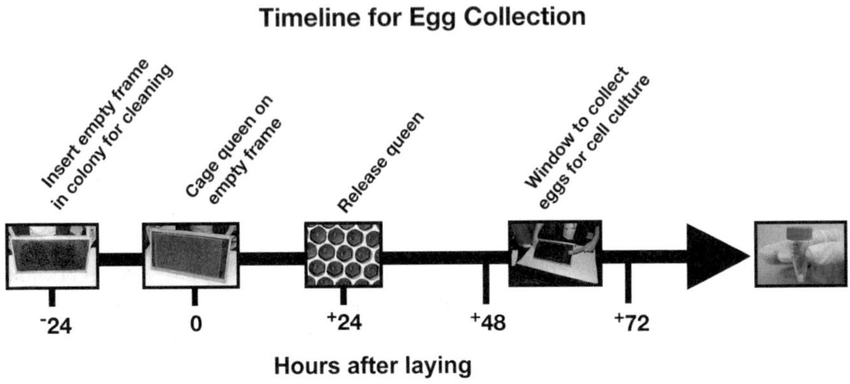

Timeline for Egg Collection

Hours after laying

Fig. 2 Timeline for the collection of honey bee eggs that are synchronized by age. Twenty-four hours before caging the queen (–24), an empty frame of drawn comb is placed in the center of the brood area of a healthy colony. Workers will clean the cells on the frame in preparation for the queen to lay eggs. The queen is then caged on one side of the frame (0) using hardware cloth with partitions that restrict her movement to the frame but are large enough to allow workers to pass freely. After 24 h, the queen is released (+24) and the frame is inspected for eggs prior to returning it to the colony for an additional 24–48 h of incubation. Between 48 and 72 h after the eggs were laid (+48 –72), most are near completion of embryogenesis. The frame containing eggs that are synchronized in their development can be brought to the laboratory and collected into a vial and processed to establish a primary cell culture

the hardware cloth are sized appropriately to allow the workers to pass freely and attend to the queen but small enough to keep the queen imprisoned. The queen is allowed to lay eggs for 24 h. She is released the following day, and the frame is inspected for her laying pattern. After the queen is released, the eggs are incubated for an additional 24–48 h. At the end of the incubation period, most eggs on the side of the frame where the queen's laying was restricted are within the window of development where there is a high rate of cell division. The frame is then transported to the laboratory where the eggs are collected into a sterile vial.

While in the vial, the eggs can be surface sterilized using disinfectants (e.g., bleach); then the eggs are rinsed several times with sterile water. A small amount of growth medium is then added to the vial, and the eggs are homogenized using a pestle to rupture the outer shell of the eggs to release the inner tissues. Once the tissues are released, they are further fragmented using the pestle. The fragments are then transferred to a flask containing medium with antibiotics. Fragments from at least ten eggs contain a density of cells sufficient to start a primary culture. Once in the flask, the tissues are then moved to an incubator set to 32 °C (89.6 °F), which is a temperature close to that of the brood nest maintained by the workers when larvae and pupae are present. This temperature has been shown to support the long-term viability of honey bee cells in culture (Hunter 2010; Goblirsch et al. 2013). Evidence for successfully initiating a primary culture can be observed as early as 3 h after transferring the tissue fragments; at this time, it should be possible to observe some cells attached to the flask substrate.

7 From a Primary Culture Grows a Cell Line

The process of adaptation for cells to their artificial environment can take weeks to months following the start of a primary culture. It is important during this time to monitor the progress of the culture and note changes to the medium or the morphology and growth rate of the cells. Exhausted growth medium is replenished with fresh medium on a regular basis, but the frequency of exchange depends on the cell density. Cultures where the cells have a higher rate of proliferation require the medium to be changed more frequently than cultures that display little change in growth rate.

There is a predictable succession in the composition of cell types over time for cells in honey bee primary cultures. Shortly after this culture is established, it is possible to see unattached fragments of tissue floating in the growth medium (Fig. 3a). If the cells in these fragments are round and translucent, it is a good indicator that they have survived the sterilization and homogenization steps. Fragments of tissue begin to attach to the flask substrate as early as 3 h after initiation of the culture. Over the course of the following days, a mosaic of cobblestone-like cells with large nuclei can be observed forming a periphery around the tissue fragments (Fig. 3b). In the weeks to months following, there is considerable outgrowth of cells away from the tissue fragments. There is also a noticeable change in the composition of the cells in the culture. The cell population begins to transition from the cobblestone-like epithelial cells to a mixed population of cells with different morphologies (Fig. 3c). Ultimately, the majority of the cells that remain in culture have a long, spindle-shaped appearance (Fig. 3d). This long, fiber-like cell type is the predominant cell type in the AmE-711 line isolated by Goblirsch et al. (2013). AmE-711 refers to the species (*Apis mellifera*), the tissue (Embryo), and the date [July (7), 2011] of the primary culture that became the continuous line.

8 AmE-711 Cells and Bee Disease

Nosema ceranae was once thought to only infect the Asian honey bee, *A. cerana* (Fries et al. 1996; Botías et al. 2012). However, a screen for microorganisms associated with the phenomenon of CCD revealed that *N. ceranae* was highly prevalent in both healthy and dying colonies of *A. mellifera* (Cox-Foster et al. 2007). Based on this study, it is plausible that this new association between *N. ceranae* and *A. mellifera* could contribute to the failing health of a honey bee colony. What is not so easy to comprehend is the fact that seemingly healthy colonies are also likely to be infected with *N. ceranae*. *Nosema ceranae* is now known to have a global distribution, as it is found on every continent where there are honey bees (Higes et al. 2006; Cox-Foster et al. 2007; Klee et al. 2007; Giersch 2009; Invernizzi et al. 2009; Higes et al. 2009). Moreover, this pathogen is capable of infecting other bee species besides *A. cerana* and *A. mellifera*, particularly

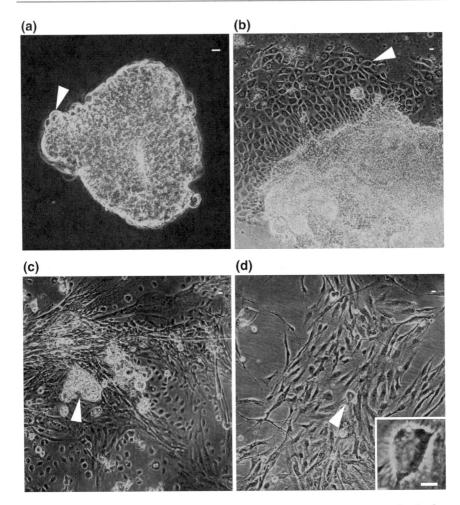

Fig. 3 The transition from primary culture to cell line. **a** Round cells (*white arrowhead*) of an unattached tissue fragment shortly after transfer to a culture flask. **b** Cobblestone-like epithelial cells with large nuclei (*white arrowhead*) growing outward from a tissue fragment attached to the flask substrate. **c** Remnants of an original tissue fragment (*white arrowhead*) 1 month after being introduced into culture. This primary culture is approaching 100% coverage of the flask substrate and contains a mixed population of different cell types; however, most notable are the long, fiber-like cells. **d** The composition of cells in the AmE-711 cell line after it had been passaged 14 times from the original primary culture. This advanced culture is composed predominantly of a single cell type, the elongate fibroblast-type cell. The white arrowhead indicates a cell that is dividing and is enlarged in the lower right corner inset. Scale bar = 10 μm. Reprinted from Goblirsch et al. 2013)

different species of bumblebee (Plischuk et al. 2009; Li et al. 2012; Chaimanee et al. 2013; Graystock et al. 2013; Fürst et al. 2014).

Without access to honey bee cells in culture, researchers have looked to other insect cells lines to develop models of infection with *N. ceranae*. Recently, lepidopteran cells lines were screened for their susceptibility to infection with *N. ceranae* (and *N. apis*). From this screen, a cell line isolated from the ovaries of the gypsy moth, *Lymantria dispar* (Goodwin et al. 1978), could be infected and permitted the development of the immature stages of *N. ceranae* (Gisder et al. 2011). Many honey bee pathogens such as *Nosema* spp. and viruses are thought to have a very limited host and tissue range that they can infect. Contrary to this paradigm, Gisder et al. (2011) demonstrated nicely that lepidopteran cells could support infection and completion of the life cycle of the bee pathogen, *N. ceranae*, under in vitro conditions.

Our laboratory followed the method used by Gisder et al. (2011) to test the susceptibility of cells from a natural host, AmE-711 cells, to infection with *N. ceranae*. We isolated spores from the guts of adult bees infected with *N. ceranae*. Purified spores were then centrifuged together with AmE-711 cells to form a pellet. A sucrose solution was added to the pellet to induce the spores to germinate and initiate the process of infection. Following germination, the AmE-711 cells were transferred to a flask where the progression of infection could be monitored. Figure 4 depicts AmE-711 cells in culture 7 days after infection with *N. ceranae* spores. From this figure, it is possible to see the different stages of development of the pathogen. For example, germinated spores appear as egg-like forms with a distinct case (E, black arrow). Germinated spores are essentially empty cases; they have evacuated their contents during the germination process, which is the mechanism of transferring their germ or reproductive material into a honey bee cell. Figure 4 also shows two different types of spores produced during *N. ceranae* infection, the immature spore (I, blue arrow), and the environmental spore (S, yellow arrow). The environmental spore is released from an infected bee cell when it dies. This is the life stage of *N. ceranae* that will pass from the gut of the bee when the bee defecates and serve as a source of infection for other bees. The environmental spore is distinguished from the immature spore by its thicker outer wall. The immature spores are tightly packed around the nucleus (N) of the bee cell. *Nosema ceranae* lacks some fundamental processes, such as the ability to create its own energy; therefore, it exploits the machinery of the bee cell for these resources. The position of the immatures spores depicted in Fig. 4 alludes to their ability to hijack host cellular machinery and rob the bee cell of its metabolic needs.

The susceptibility of the AmE-711 cell line to infection with *N. ceranae* opens the door to understanding how this pathogen gains access to and exploits its host. For example, there remain gaps in our knowledge of how a spore finds its target during the germination process. A culture system of bee cells could also aid in identifying factors that contribute to the virulence of different strains of *N. ceranae*. Further, molecular techniques could be used to compare the response of a cell's micro-managers (i.e., the expression profile of its genes and proteins) to infection with different strains of *N. ceranae*, leading to the identification of markers of

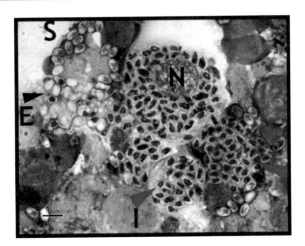

Fig. 4 Infection of AmE-711 cells with *Nosema ceranae*. Spores of *N. ceranae* were germinated in the presence of AmE-711 cells. Immature spores are denoted by I, *blue arrows*. The immature spores are packed tightly around the nucleus (N) of the honey bee cell. Immatures spores have a thin spore wall compared to mature spores; the thin wall allows the stain to penetrate more easily into the spore interior, giving immature spores a *darker violet* appearance. Mature spores are marked with S, *yellow arrow*. The presence of the different spore types demonstrates that *N. ceranae* can complete its life cycle, from germination to the production of mature spores, within honey bee cells in culture. Germinated spores (E, *black arrow*) are differentiated from environmental and immatures spores because they lack definition within the egg-like case. The germinated spore is an empty spore case that has evacuated its contents during the germination process. Scale bar = 10 μm

resistance or tolerance. This information could be applied toward the development of detection assays that bee breeders could use for the selection of resistant stocks of bees.

9   Persistent Infection of AmE-711 with a Honey Bee Virus

Honey bees are host to numerous viruses, and simultaneous infection with multiple viruses is common. Viruses frequently detected in colonies include black queen cell virus (BQCV), deformed wing virus (DWV), Kashmir bee virus (KBV), and sacbrood virus (SBV) (Chen and Siede 2007; Welch et al. 2009). Like *N. ceranae*, bee viruses are able to infect many species of bumblebee (Genersch et al. 2006; Meeus et al. 2011; Kojima et al. 2011; Peng et al. 2011), as well as other non-hymenopteran insects (Singh et al. 2010; Levitt et al. 2013). Most viruses that infect honey bees use a single, relatively short strand of RNA as their genome. Typically, viruses persist as asymptomatic infections in bees until activated by certain stressors, such as parasitism by *Varroa* mites. One virus, in particular, DWV, is highly prevalent in honey bee colonies and is distributed across a wide geographic range (Traynor et al. 2016). Adult bees that are symptomatic for DWV have non-functional wings and a shortened life span. Moreover, DWV is thought to be the main contributor to the high overwintering mortality that many beekeepers experience (Highfield et al. 2009; Dainat et al. 2012).

Due to the impact that DWV and other viruses have on honey bee health, there is a pressing need to expand our understanding of triggers that cause these pathogens to transition from asymptomatic to acute infections. Recently, it was shown that the AmE-711cell line was persistently infected with DWV (Carrillo-Tripp et al. 2016). This finding provides the opportunity to study the virus–host interaction of DWV at its most basic level. Although infection of AmE-711 cells with DWV cannot be observed directly without high-powered magnification, it is possible to observe cytopathic effects caused by the virus, including rounding and detachment of infected cells from the substrate, blebbing or protrusions of the cell membrane, and rupturing of cells. The viral load in dying versus seemingly healthy cells can also be quantified using molecular techniques. These capabilities allow researchers to expose AmE-711 cells to different stressors such as temperature change and observe the response of DWV, such as an increase in viral load and cell death. These conditions could then be applied to bees in cages or colonies to see whether they trigger an onset of acute infection and what is the subsequent response of the bee or colony. Although there are no current therapeutics that are effective for use against honey bee viruses, gene editing technologies are advancing that they may soon be able to modify the genome of the bee or the virus in a way that prevents viral entry or replication. The AmE-711 cell line or other honey bee culture system could be instrumental in testing the efficiency and safety of these novel approaches.

10 Uses for Other Honey Bee Cell Culture Systems

10.1 Honey Bee Cells for Toxicology Studies

The AmE-711 cell line was isolated from undifferentiated embryonic tissues. Other honey bee cell lines composed of specific cell types would be useful for research directed at specific tissues. For example, if we had a cell line derived from nervous tissue (i.e., a neuron-derived culture system), we could apply it to study the effects of environmental pesticides on the nervous system of honey bees. Neonicotinoids, a class of insecticides that act much like nicotine in the human brain by overstimulating neurons, are considered to be a contributing factor to honey bee decline. These insecticides are applied to crops and ornamental plants for the control of pest species. At high doses, neonicotinoids cause death and paralysis in insects. Honey bees are not the intended targets of neonicotinoid applications but may be exposed during foraging for pollen and nectar on treated plants or plants contaminated from spray drift. Honey bee foragers exposed to sublethal doses of neonicotinoids exhibit behaviors that suggest a disruption of motor and cognitive functioning (Blacquière et al. 2012). Exposing honey bee neuronal cells to different doses of neonicotinoids would make it possible to more thoroughly characterize the molecular mechanisms responsible for this intoxication-like state in exposed foragers (Blacquière et al. 2012). Moreover, information on the interaction of neonicotinoid toxins with receptors on neurons that respond to these toxins could be used to develop pesticides with greater efficacy/specificity against target insects while posing minimal risk to bees and other beneficial insects. For example, molecular techniques could be used to modify existing, or create novel, toxins that have high binding affinity with receptors of the pest insect but low to no binding affinity with the honey bee. This binding specificity could be tested using neuronal cells from the honey bee and from the pest species, where changes in cell viability are measured as a response to exposure to the toxin. It is unfortunate that there are no insect neuronal cell lines. However, this does not preclude the use of primary cultures containing honey bee neurons toward answering tissue-specific questions about the honey bee nervous system.

10.2   Honey Bee Midgut Cell Cultures

A cell line composed of a single cell type allows access to a homogenous, replicable pool of biological material that can be maintained under precisely controlled conditions. One tissue that has proven successful as a donor tissue to establish insect cell lines is the midgut (Hakim et al. 2009). In honey bees, cells that line the midgut have a limited life span due to damage caused by abrasions from pollen digestion and detoxification of chemicals such as pesticides. Housed within the honey bee midgut are so-called intestinal crypts, pockets of dividing stem cells tasked with replacing damaged and dying cells (Ward et al. 2008). Recently, Zhang et al. (2009) demonstrated the potential of cell cultures established from honey bee midgut tissues by maintaining cells from *A. cerana* larvae for 5 months. Other reports on honey bee midgut cell culture have not been forthcoming. One explanation for the lack of honey bee midgut cell cultures may stem from the effort needed to remove the digestive system concealed within the intact bee under sterile conditions. There is a high probability that a primary culture of bee midgut cells will become contaminated with microbes that naturally inhabit the digestive system. The midgut and adjacent tissues (i.e., crop, ileum, and rectum) are a microcosm of bacteria, fungi, protozoa, and viruses. Adding antibiotics to the medium can delay the spread of contamination. Antibiotics limit the ability of microorganisms to reproduce, but their overuse can also put microbes under strong selection pressure. Virulent strains that survive the selection pressure may overgrow a culture of cells when the agents are removed from the medium.

10.3 Honey Bee Cell Culture to Help Ensure the Future of Honey Bees

The interaction of pathogens, mite pests, pesticides, and poor nutrition is likely the driving force behind declining health of honey bee colonies in the USA and other parts of the world. To tackle this complex issue, beekeepers and researchers need novel approaches to complement traditional methodologies. Advances in human, animal, and plant health have benefitted greatly from the use of cell lines to study factors that cause disease in these systems, especially pathogens. For this reason, cell lines established from honey bee tissues can be used to study honey bee biology and disease in a simplified host environment. The AmE-711 cell line and other in vitro culture systems derived from honey bee tissues have the potential to impact our understanding of how pathogens interact with their host at the cellular and molecular level to negatively affect honey bee health. Honey bee culture systems are not limited to studying the causes of honey bee disease, but could be applied more broadly to areas of developmental biology, neurobiology, and functional genomics. Honey bee cell culture systems, such as the AmE-711 cell line, are a means of *working small to solve big.*

With all of the specialized equipment and training needed for this work, it is truly not an area that can be of direct use to the citizen scientist or beekeeper. However, ultimately, the use and development of honey bee cell lines can be expected to result in expanding our knowledge on detractors to honey bee health. Understanding mechanisms of bee disease at the cell level will spur efforts between beekeepers and scientists to develop approaches that integrate bee nutrition, management practices, and novel therapeutics that sustain populations of healthy and productive bees.

Acknowledgements I am indebted to Drs. Marla Spivak and Timothy Kurtti for mentoring me through the trials and tribulations of establishing a honey bee cell line. I am also grateful for the valuable feedback they, Ana Heck, and Dr. Russell Vreeland, provided on an earlier version of this manuscript.

References

Airenne KJ, Hu YC, Kost TA, Smith RH, Kotin RM, Ono C, Matsuura Y, Wang S, Ylä Herttuala S (2013) Baculovirus: an insect-derived vector for diverse gene transfer applications. Mol Ther 21(4):739–749

Arif B, Pavlik L (2013) Insect cell culture: virus replication and application in biotechnology. J Invertebr Pathol 112:S138–S141

Barbara GS, Grünewald B, Paute S, Gauthier M, Raymond-Delpech V (2008) Study of nicotinic acetylcholine receptors on cultured antennal lobe neurons from adult honeybee brains. Invert Neurosci 8(1):19–29

Beisser K, Munz E, Reimann M, Renner-Müller IC (1990) Experimental studies of *in vitro* cultivation of the cells of Kärtner honeybees (*Apis mellifera carnica* Pollmann, 1879). Zentralbl Veterinarmed B 37(7):509–519

Bergem M, Norberg K, Aamodt RM (2006) Long-term maintenance of *in vitro* cultured honeybee (*Apis mellifera*) embryonic cells. BMC Dev Biol 6:17

Bicker G, Kreissl S (1994) Calcium imaging reveals nicotinic acetylcholine receptors on cultured mushroom body neurons. J Neurophysiol 71(2):808–810

Blacquière T, Smagghe G, van Gestel CAM, Mommaerts V (2012) Neonicotinoids in bees: a review on concentrations, side-effects and risk assessment. Ecotoxicology 21(4):973–992

Botías C, Anderson DL, Meana A, Garrido-Bailón E, Martín-Hernández R, Higes M (2012) Further evidence of an oriental origin for *Nosema ceranae* (Microsporidia: Nosematidae). J Invertebr Pathol 110(1):108–113

Carrillo-Tripp J, Dolezal AG, Goblirsch MJ, Miller WA, Toth AL, Bonning BC (2016) *In vivo* and *in vitro* infection dynamics of honey bee viruses. Sci Rep 6:22265

Chaimanee V, Pettis JS, Chen Y, Evans JD, Khongphinitbunjong K, Chantawannakul P (2013) Susceptibility of four different honey bee species to *Nosema ceranae*. Vet Parasitol 193(1–3): 260–265

Chan MMY, Choi SYC, Chan QWT, Li P, Guarna MM, Foster LJ (2010) Proteome profile and lentiviral transduction of cultured honey bee (*Apis mellifera* L.) cells. Insect Mol Biol 19(5):653–658

Chen YP, Siede R (2007) Honey bee viruses. Adv Virus Res 70:33–80

Cox-Foster DL, Conlan S, Holmes EC et al (2007) Science 318(5848):283–287

Dainat B, Evans JD, Chen YP, Gauthier L, Neumann P (2012) Dead or alive: deformed wing virus and Varroa destructor reduce the life span of winter honeybees. Appl Environ Microbiol 78:981–987

DuPraw EJ (1967) The honeybee embryo. In: Wilt FH, Wessells NK, Wessells NK (eds) Methods in developmental biology. Cromwell, New York, p 183–217

Fries I, Feng F, da Silva A et al. (1996) *Nosema ceranae* n. sp. (Microspora, Nosematidae), morphological and molecular characterization of a microsporidian parasite of the Asian honey bee *Apis cerana* (Hymenoptera, Apidae). Eur J Protistol 32(3):356–365

Fürst MA, McMahon DP, Osborne JL, Paxton RJ, Brown MJ (2014) Disease associations between honeybees and bumblebees as a threat to wild pollinators. Nature 506(7488):364–366

Gascuel J, Masson C, Bermudez I, Beadle DJ (1994) Morphological analysis of honeybee antennal cells growing in primary cultures. Tissue Cell 26(4):551–558

Genersch E, Yue C, Fries I, de Miranda JR (2006) Detection of deformed wing virus, a honey bee viral pathogen, in bumble bees (*Bombus terrestris* and *Bombus pascuorum*) with wing deformities. J Invertebr Pathol 91(1):61–63

Giauffret A (1971) Cell culture of Hymenoptera. In: C Vago (ed) Invertebrate tissue culture: volume 2. Academic Press, New York, 295–305

Giauffret A, Quiot JM, Vago C, Poutier F (1967) *In vitro* culture of cells of the bee. C R Acad Sci Hebd Seances Acad Sci D. 265:800–803

Giersch T, Berg T, Galea F, Hornitzky M (2009) *Nosema ceranae* infects honey bees (*Apis mellifera*) and contaminates honey in Australia. Apidologie 40(2):117–123

Gisder S, Möckel N, Andreas L, Genersch E (2011) A cell culture model for *Nosema ceranae* and *Nosema apis* allows new insights into the life cycle of these important honey bee-pathogenic microsporidia. Environ Micrbiol 13(2):404–413

Gisselmann G, Warnstedt M, Gamerschlag B et al (2003) Characterization of recombinant and native Ih-channels from *Apis mellifera*. Insect Biochem Mol Biol 33(11):1123–1134

Goblirsch MJ, Spivak MS, Kurtti TJ (2013) A cell line resource derived from honey bee (*Apis mellifera*) embryonic tissues. PLoS ONE 8(7):e69831

Goldberg F, Grünewald B, Rosenboom H, Menzel R (1999) Nicotinic acetylcholine currents of cultured Kenyon cells from the mushroom bodies of the honey bee *Apis mellifera*. J Physiol 514(3):759–768

Goldschmidt R (1915) Some experiments on spermatogenesis *in vitro*. Proc Natl Acad Sci USA 1(4): 220–222

Goodwin RH, Tompkins GJ, McCawley P (1978) Gypsy moth cell lines divergent in viral susceptibility. In Vitro 14:485–494

Grace TDC (1962) Establishment of four strains of cells from insect tissues grown in vitro. Nature 195:788–789

Graystock P, Yates K, Darvill B, Goulson D, Hughes WO (2013) Emerging dangers: deadly effects of an emergent parasite in a new pollinator host. J Invertebr Pathol 114(2):114–119

Grünewald B (2003) Differential expression of voltage-sensitive K + and Ca2 + currents in neurons of the honeybee olfactory pathway. J Exp Biol 206(1):117–129

Hakim RS, Caccia S, Loeb M, Smagghe G (2009) Primary culture of insect midgut cells. Vitro Cell Dev Biol Anim 45(3–4):106–110

Higes M, Martín R, Meana A (2006) Nosema ceranae, a new microsporidian parasite in honeybees in Europe. J Invertebr Pathol 92(2):93–95

Higes M, Martín-Hernández R, Garrido-Bailón E, Botías C, Meana A (2009) The presence of Nosema ceranae (Microsporidia) in North African honey bees (Apis mellifera intermissa). J Apicult Res 48(3):217–219

Highfield AC, El Nagar A, Mackinder LC, Nöel LM, Hall MJ, Martin SJ, Schroeder DC (2009) Deformed wing virus implicated in overwintering honeybee colony losses. Appl Environ Microbiol 75:7212–7220

Hunter WB (2010) Medium for development of bee cell cultures (Apis mellifera: Hymenoptera: Apidae). Vitro Cell Dev Biol Anim 46(2):83–86

Invernizzi C, Abud C, Tomasco IH, Harriet J, Ramallo G, Campá J, Katz H, Gardiol G, Mendoza Y (2009) Presence of Nosema ceranae in honeybees (Apis mellifera) in Uruguay. J Invertebr Pathol 101(2):150–153

Ju H, Ghil S (2015) Primary cell culture method for the honeybee Apis mellifera. Vitro Cell Dev Biol Anim 51(9):890–893

Kitagishi Y, Okumura N, Yoshida H, Nishimura Y, Takahashi J, Matsuda S (2011) Long-term cultivation of in vitro Apis mellifera cells by gene transfer of human c-myc proto-oncogene. Vitro Cell Dev Bio Anim 47(7):451–453

Klee J, Besana AM, Genersch E, Gisder S, Nanetti A, Tam DQ, Chinh TX, Puerta F, Ruz JM, Kryger P, Message D, Hatjina F, Korpela S, Fries I, Paxton RJ (2007) Widespread dispersal of the microsporidian Nosema ceranae, an emergent pathogen of the western honey bee Apis mellifera. J Invertebr Pathol 96(1):1–10

Kloppenburg P, Kirchhof BS, Mercer AR (1999) Voltage-activated currents from adult honeybee (Apis mellifera) antennal motor neurons recorded in vitro and in situ. J Neurophysiol 81(1):39–48

Kojima Y, Toki T, Morimoto T, Yoshiyama M, Kimura K, Kadowaki T (2011) Infestation of Japanese native honey bees by tracheal mite and virus from non-native European honey bees in Japan. Microb Ecol 62(4):895–906

Kreissl S, Bicker G (1992) Dissociated neurons of the pupal honeybee brain in cell culture. J Neurocytol 21(8):545–556

Levitt AL, Singh R, Cox-Foster DL et al (2013) Cross-species transmission of honey bee viruses in associated arthropods. Virus Res 176(1–2):232–40

Li J, Chen W, Wu J, Peng W, An J, Schmid-Hempel P, Schmid-Hempel R (2012) Diversity of Nosema associated with bumblebees (Bombus spp.) from China. Int J Parasitol 42(1):49–61

Lynn DE (2003) Novel techniques to establish new insect cell lines. Vitro Cell Dev Biol Anim. 37 (6):319–321

Malun D, Moseleit AD, Grünewald B (2003) 20-Hydroxyecdysone inhibits the mitotic activity of neuronal precursors in the developing mushroom bodies of the honeybee Apis mellifera. J Neurobiol 57(1):1–14

Meeus I, Brown MJ, De Graaf DC, Smagghe G (2011) Effects of invasive parasites on bumble bee declines. Conserv Biol 25(4):662–671

Peng W, Li J, Boncristiani H, Strange J, Hamilton M, Chen Y (2011) Host range expansion of honey bee black queen cell virus in the bumble bee, Bombus huntii. Apidologie 42(5):650–658

Plischuk S, Martín-Hernández R, Prieto L, Lucía M, Botías C, Meana A, Abrahamovich AH, Lange C, Higes M (2009) South American native bumblebees (Hymenoptera: Apidae) infected by *Nosema ceranae* (Microsporidia), an emerging pathogen of honeybees (*Apis mellifera*). Environ Microbiol Rep 1(2):131–135

Pollinator Health Task Force (2015) National strategy to promote the health of honey bee and other pollinators. Available at https://www.whitehouse.gov/sites/default/files/microsites/ostp/Pollinator%20Health%20Strategy%202015.pdf

Schäfer S, Rosenboom H, Menzel R (1994) Ionic currents of Kenyon cells from the mushroom body of the honeybee. J Neurosci 14(8):4600–4612

Singh R, Levitt AL, Rajotte EG, Holmes EC, Ostiguy N, Vanengelsdorp D, Lipkin WI, Depamphilis CW, Toth AL, Cox-Foster DL (2010) RNA viruses in hymenopteran pollinators: evidence of inter-taxa virus transmission via pollen and potential impact on non-*Apis* hymenopteran species. PLoS ONE 5(12):e14357

Smith GE, Ju G, Ericson BL, Moschera J, Lahm HW, Chizzonite R, Summers MD (1985) Modification and secretion of human interleukin 2 produced in insect cells by a baculovirus expression vector. Proc Natl Acad Sci USA 82(24):8404–8408

Stanley M (1968) Initial results of honeybee tissue culture. Bull Apic 11:45–55

Traynor KS, Rennich K, Forsgren E, Rose R, Pettis J, Kunkel G, Madella S, Evans J, Lopez D, vanEngelsdorp D (2016) Multiyear survey targeting disease incidence in US honey bees. Apidologie 47:325–348

Walker T, Jeffries CL, Mansfield KL, Johnson N (2014) Mosquito cell lines: history, isolation, availability and application to assess the threat of arboviral transmission in the United Kingdom. Parasit Vectors 7:382

Ward KN, Coleman JL, Clinnin K, Fahrbach S, Rueppell O (2008) Age, caste, and behavior determine the replicative activity of intestinal stem cells in honeybees (*Apis mellifera* L.). Exp Gerontol 43(6):530–537

Welch A, Drummond F, Tewari S, Averill A, Burand JP (2009) Presence and prevalence of viruses in local and migratory honeybees (*Apis mellifera*) in Massachusetts. Appl Environ Microbiol 75(24):7862–7865

Zhang G-Z, Shuang T, Yi Z, Han R-C (2009) Establishment and maintenance of *in vitro* cultured Chinese honeybee *Apis ceranae* midgut epithelial cells. Sociobiology 54(1):5–18

Author Biography

Michael Goblirsch earned his Ph.D. from the Department of Entomology at the University of Minnesota. The focus of his dissertation was to understand the role of the fungal pathogen, *Nosema ceranae*, in honey bee health. His current research focuses on the development of honey bee cell culture systems to further understanding of intracellular microbes that have a negative impact on bee health, including the many bee viruses. Mike is a hobby beekeeper and enjoys working with and especially learning from other beekeepers. He is currently a postdoctoral research associate in the Bee Lab at the University of Minnesota.

Honey Bee Viruses—Pathogenesis, Mechanistic Insights, and Possible Management Projections

Nor Chejanovsky and Yossi Slabezki

Abstract

Honey bee viruses have gained substantial attention due to their involvement in the collapse of honey bee colonies. This chapter focuses on honey bee viruses linked to honey bee colony losses, specifically those that cause paralysis, those carried by Varroa mites, and those that cause deformed wings. Often virus infections in the colony are dormant and asymptomatic. Asymptomatic infections can convert to active (and visible) symptomatic infections when colonies are exposed to various stresses. These stresses include biological, such as *Varroa destructor*, mechanical, such as the utilization of bee colonies for pollination in net-covered crops, and chemical, such as the use of insecticides harmful to bees. These stresses enable viruses to overcome natural honey bee defenses, by facilitating viral access to the bee blood (hemolymph) and by weakening its immune system. Knowledge and understanding of the cause-and-effect interactions between viruses, stress factors, and honey bees will promote the use of antistress measures to help ameliorate collapse of honey bee colonies. This chapter is the result of intense collaboration between Y.S., instructor in beekeeping for the Extension Service of the Ministry of Agriculture and N.C., researcher of insect viruses and particularly honey bee viruses at ARO. The subjects presented below try to integrate the beekeeping and virus pathology perspectives.

N. Chejanovsky (✉)
Department of Entomology, Institute of Plant Protection, The Volcani Center,
Bet Dagan 50250, Israel
e-mail: ninar@volcani.agri.gov.il

Y. Slabezki
Beekeeping Division, Extension Service Israeli Ministry of Agriculture,
Bet Dagan 50250, Israel
e-mail: yoslav@shaham.moag.gov.il

© Springer International Publishing AG 2017   109
R.H. Vreeland and D. Sammataro (eds.), *Beekeeping – From Science to Practice*,
DOI 10.1007/978-3-319-60637-8_7

1 Honey Bee Viruses and Colony Losses

Honey bee viruses have gained substantial attention since the first reports of colony collapse disorder (CCD) where many honey bee (*Apis mellifera*) colonies were lost in the US during 2006–2007 (Cox-Foster et al. 2007; Stokstad 2007). As a result, it became clear to a wider public that bees were in trouble. Several pathogenic viruses were then found to be actively involved in the collapse of honey bee colonies around the world (Cox-Foster et al. 2007; Berthoud et al. 2013; Chen and Siede 2007; Cornman et al. 2012; Gensersch et al. 2010; van Engelsdorp et al. 2009).

The most common honey bee viruses currently recognized are acute bee paralysis virus (ABPV), black queen cell virus (BQCV), chronic bee paralysis virus (CBPV), deformed wing virus (DWV), Israeli acute bee paralysis (IAPV), Kashmir bee virus (KBV), sacbrood virus (SBV), and Varroa destructor-1 (VDV-1) (Chen and Siede 2007; de Miranda et al. 2010 2013; de Miranda and Gensersch 2010; Ribiere et al. 2010) (see Table 1).

Table 1 Honey bee viruses discussed in this chapter and their symptoms

Virus name	Abbreviation	Clade and family	Symptoms	Transmission by *V. destructor*
Acute bee paralysis virus	ABPV	ABPV-IAPV-KBV *Dicistroviridae*	Paralysis including: trembling, leg paralysis, the inability to fly, and general paralysis that leads to death. No dead bees accumulate in front of the colony	Yes
Kashmir bee virus	KBV			
Israeli acute bee paralysis	IAPV			
Chronic bee paralysis virus	CBPV	Unclassified	Paralysis involving abnormal trembling of body and wings. Inability to fly, crawling at the beehive entrance and on the ground. Bloated abdomens and hairless bees with black coloration on the abdomen. Piles of dead bees accumulate in front of the colony.	No
Deformed wing virus	DWV	DWV-VDV-1-KV *Iflaviridae*	Deformed wings, bloated and shortened abdomens, discoloration, and premature death	Yes
*Varroa destructor-*1	VDV-1			
Kakugo virus	KV			

Renewed research has identified new viruses infectious to honey bees, such as the various strains of Lake Sinai virus and aphid lethal paralysis virus Brookings strain (Runckel et al. 2011, Cornman et al. 2012). The real impact on honey bee colonies of some of the latter viruses is still unknown [for a comprehensive list of viruses, see (de Miranda et al. 2013, Runckel et al. 2011)].

In this chapter, we will focus on the major honey bee viruses responsible for recent colony losses. We will distinguish between viruses that cause paralysis (acute or chronic) and those that cause the easily recognized symptom of emerging bees with deformed wings (see Table 1).

2 Virus-Mediated Paralysis of the Honey Bee

The paralysis group of viruses Table 1 (de Miranda et al. 2010; Ribiere et al. 2010) may be present in the colonies in covert asymptomatic infections (not visible) with the symptoms described below appearing after the virus progresses to a more virulent form.

ABPV-IAPV-KBV belong to the viral family *Dicistroviridae*, due to the nature and specific organization of the viral genome; a single-strand RNA molecule bearing all the information the virus needs to replicate in the cells of its host (de Miranda et al. 2010).

Symptoms associated with paralysis viruses include trembling, leg paralysis, the inability to fly, and general paralysis that leads to death [most often observed for ABPV and IAPV and less frequently for KBV infections (de Miranda et al. 2010)]. In IAPV-infected hives, a relatively high number of smaller bees are often seen. Some researchers reported dark cuticle pigmentation in adult bees infected with IAPV and in pupae experimentally injected with the virus (Boncristiani et al. 2013; Maori et al. 2007), but in experiments performed with emerging bees fed with highly purified viral stocks, the bees never showed this symptom (Y.S. and N.C., unpublished observations). Paralysis by this group of viruses does not seem to result in accumulation of dead bees in front of the beehive (de Miranda et al. 2010).

IAPV was initially linked to CCD because CCD colonies had high loads of this infectious virus (Cox-Foster et al. 2007; Hou et al. 2014). Interestingly, IAPV was detected in the heads of experimentally infected foragers that showed impaired cognition and homing ability (Li et al. 2013). A recent study showed that IAPV was most abundant in the gut, hypopharyngeal glands, and the nerves of infected adults (Chen et al. 2014). Queens can bear the virus in the gut, spermatheca, and ovary and can lay infected eggs as well (Chen et al. 2014). Newly emerging bees are very sensitive to oral infection (mostly by trophallaxis).

The ectoparasite *Varroa destructor* is able to transmit viruses of this family, though it seems that this happens less frequently than transmission of viruses of the DWV clade (see below). In the USA and Europe, ABPV and IAPV prevalence increases in the summer (Bailey et al. 1981; de Miranda et al. 2010; Chen et al. 2014), while in Israel, its prevalence peaked mostly in the fall (Soroker et al. 2011).

CCD colonies detected in Israel had active IAPV infection with higher viral loads in April and December (Hou et al. 2014). Acute paralysis virus (ABPV) was discovered as a contaminant CBPV viral stocks (Bailey et al. 1963).

CBPV displays a different genomic organization (two single RNA segments of different size packaged in the viral particles) and is not classified in any viral family yet (Ribiere et al. 2010). From sequence analysis (the nature of the genomic information), Lake Sinai viruses display partial similarity to CBPV; however, no specific symptoms were associated with their infections in honey bees (Runckel et al. 2011).

CBPV paralysis involves abnormal trembling of the body and wings. Symptomatic bees are not able to fly and often crawl at the beehive entrance and on the ground, and piles of dead bees can be seen in front of the colony. Bloated abdomens and hairless bees with black coloration on the abdomens were also detected (Ribiere et al. 2010). The virus seems not to be transmitted by Varroa mites (*V. destructor*) (Ribiere et al. 2010). The infection develops slowly, from 6 days to two weeks, depending upon the conditions and probably the viral strain. Following CBPV infections in Israel, we were able to distinguish two types of infections:

1. an individual infected colony shows the typical symptoms of paralysis that are usually detected by the end of the winter and beginning of the spring and,
2. a group of colonies become infected and dead bees pile up in front of the colony during the spring-to-summer transition seasons.

Recently, we found that in most cases, CBPV infections were accompanied by ABPV infections. We are currently investigating whether the type of infections presented above have any correlation with the amount of ABPV present in single-versus group-type infected colonies.

What factors determine the type of infection? It could be the evolution of the virus to more virulent/infective strains, environmental interactions difficult to reveal, and even characteristics of the colony. Research is ongoing to answer this question. CBPV was also reported to be able to prevail in the colony in an asymptomatic state (Ribiere et al. 2010).

CBPV exhibits broad distribution in the infected bee; remarkably high numbers of viral particles were detected in the head. CBPV also prefers the honey bee nervous system. Also, high numbers of viral copies (around 10^9 per μl) were detected in the hemolymph of the infected host. The high preference of CBPV for the bee's nervous system correlates with trembling and other typical paralysis symptoms observed in adult bees from infected hives (Ribiere et al. 2010).

CBPV infects adults, brood, and also eggs, but the virus replicates to higher titers in worker bees (Blanchard et al. 2007). Experimental infections showed that honey bee queens are susceptible to CBPV, probably transmitted by trophallaxis. However, in naturally infected hives, there seem to exist behavioral strategies that prevent the queen from being fed by infected workers (Amiri et al. 2014). Also, CBPV can be transmitted by contact between infected bees and their non-infected mates, as well as by oral ingestion of infected feces that have high viral loads (Ribiere et al. 2007).

CBPV was sometimes reported in association with *Nosema ceranae* infections (Toplak et al. 2013).

3 Deformed Wing Virus Clade

In this group, we find the viruses of the DWV-VDV-1-Kakugo virus (KV) clade (de Miranda and Genersch 2010). Though KV was mainly associated with aggressive behavior of infected bees and VDV-1 was initially found in Varroa mites, this group forms part of the clade because of the similarity of their genomes with DWV (de Miranda and Genersch 2010; Fujiyuki et al. 2005; Ongus et al. 2004). DWV-VDV-1 and KV belong to the *Iflaviridae* family of viruses [also with a single-stranded RNA molecule similar to the dicistroviruses, but displaying a different organization (de Miranda and Genersch 2010)].

Queens, workers, and brood can be infected with viruses of the DWV clade (de Miranda and Genersch 2010). Vertical transmission by drones and queens was reported as well (Fievet et al. 2006). Horizontal transmission by larval food and trophallaxis was also reported; however, the oral route of infection mostly results in asymptomatic infections. Before the invasion of Varroa, DWV was often present in honey bee colonies as an asymptomatic or mild infection (Gauthier et al. 2007; de Miranda and Genersch 2010). The spread of *V. destructor* throughout the world contributed to the horizontal transmission of DWV, mostly by the ability of the mite to carry and inject the virus directly into the bee hemolymph. This direct injection promotes the conversion of avirulent or low virulent asymptomatic viruses to more virulent viruses that induced symptomatic infections [(Moore et al. 2011; Ryabov et al. 2014) and see Sect. 4.1]. DWV and VDV-1 were shown to replicate in the mite as well but they seem not to harm it (Ongus et al. 2004; Shen et al. 2005; Yue and Genersch 2005; Tentcheva et al. 2006).

Worker honey bees infected with virulent DWV/VDV-1-like viruses displayed wing deformation, bloated and shortened abdomens, and discoloration (de Miranda and Genersch 2010, Zioni et al. 2011, de Miranda et al. 2013) and resulted in premature death of the bees.

DWV has been detected in the midgut of infected workers (Fievet et al. 2006) and in the hemolymph of Varroa-parasitized individuals as well as in the gut, wings, legs, head, thorax, and abdomen (Boncristiani et al. 2009; Shah et al. 2009). High loads of virus were also localized to the heads of infected workers (Yue and Genersch 2005; Zioni et al. 2011). Interestingly, DWV-infected bees showed learning disabilities (Iqbal and Mueller 2007). Moreover, extremely virulent strains may cause premature death of infected larvae parasitized with Varroa, aborting the emergence of worker bees (Martin 2001). The increasing imbalance in the bee population composition in such infected colonies can lead to their subsequent collapse (Dainat et al. 2012a). DWV has a worldwide distribution and in Europe and Israel, it is the most prevalent virus (Genersch et al. 2010; Soroker et al. 2011; de Miranda and Genersch 2010; Berthoud et al. 2013). In Europe, DWV is highly associated with losses of overwintering colonies (Dainat et al. 2012b; Highfield et al. 2009).

4 An Abrupt Awakening: Stress-Induced Viral Infections

As discussed above, honey bee viruses can be carried by individual bees in an asymptomatic or silent mode. This equilibrium between the host and the pathogen can be broken by the appearance of outside stress factors, such as chemical or biological stresses that can induce replication of dormant viruses. In this section, scenarios of biological, chemical, and other stresses that may cause dormant viruses to replicate and cause symptoms will be covered.

4.1 DWV and the Biological Vector *Varroa destructor*, a Vicious Cycle

The rapid expansion of the ectoparasite *V. destructor* throughout the globe from the Eastern honey bee *A. cerana* to the Western honey bee *A. mellifera* introduced a new stress factor to Western bee colonies since the viruses were mostly asymptomatic (de Miranda and Genersch 2010). Varroa serves as a vector of viruses, thus profoundly changing the manner of transmission (Yue and Genersch 2005). Also, several investigations indicated that Varroa exerts a debilitating immunosuppressive effect in the parasitized bee (Shen et al. 2005; Nazzi et al. 2012). DWV became one of the most prevalent viruses in honey bee colonies and collapsing colonies showing typical symptoms of DWV infections became more frequent (de Miranda and Genersch 2010). Furthermore, the number of Varroa mites that could induce the collapse of a colony at the beginning of the Varroa invasion diminished over time. For example, in Germany at the beginning of Varroa infestation of *A. mellifera,* the colonies were able to sustain high levels of mites, up to 10,000, but nowadays mite levels above 3000 may be enough to cause colony collapse (Boecking and Genersch 2008). Varroa-parasitized bees with deformed wings symptoms showed very high loads of DWV-like viruses (Gisder et al. 2009; Zioni et al. 2011).

During the beekeeping season, when the colonies display high brood activity and rapid population increase (due to the abundant forage), no treatment against Varroa is usually applied to avoid contaminating the honey with chemicals. Thus, the Varroa population increases and concomitantly the DWV-like viruses, which is often unnoticed. But when Varroa treatments begin, the viruses do not necessarily disappear. Harsher climatic conditions, like the European winter or warm Middle Eastern summer, when forage is poor, lead to shortened life span of virus-infected adults and rapid bee depopulation of the colony. Since the colony is unable to replace the lost bees with a strong buildup of younger bees. Thus, despite success in controlling/combating Varroa, the colonies may collapse with characteristic post-Varroa syndrome.

An insight to the nature of this phenomenon was explained in various studies. Ongus et al. (2004) discovered VDV-1 that appeared to replicate in the mite. VDV-1 is highly homologous to DWV [about 84% similarity at the genomic level (Ongus et al. 2004)]. In addition, DWV replicates to high loads in the mite and mostly in the head of symptomatic bees (Gisder et al. 2009; Yue and Genersch 2005). Moreover, DWV-symptomatic bees bore recombinant DWV/VDV-1 viruses in their heads (Moore et al. 2011; Zioni et al. 2011). These results suggested that parasitism by Varroa provoked not only a significant increase in viral prevalence and a quantitative change enhancing replication of DWV, but also a qualitative change in the virus, selecting from a mild to a more virulent strain. These hypotheses were confirmed by two studies: one following the invasion of Varroa to the Hawaiian Islands under natural conditions and the second in the UK with experimentally infected hives (Martin et al. 2012; Ryabov et al. 2014). This suggested that either the immunosuppressing activity of Varroa on the honey bee and/or the ability of the virus to replicate to high loads in the mite and in the bee promoted the transformation of DWV and the appearance of DWV-VDV-1 recombinants and virulent DWV strains (Martin et al. 2012; Ryabov et al. 2014).

Nazzi et al. (2012) demonstrated that Varroa and DWV build on weakening of the bee's immune system mediated by NfκB, a protein that regulates its stress-related responses (Nazzi et al. 2012). At high DWV loads (over 10^{15} viral copies per bee), this results in the down-regulation of genes involved in the immune response of the honey bee. Thus, the renewed ability of the virus to change (mediated by Varroa) and replicate to higher loads could benefit the parasite, whose gain could be a reduction in the ability of the bee host to react to it (e.g. to being wounded which is known to trigger immune responses (Nazzi et al. 2012)]. What is the advantage to the virus? Further studies showed that direct injection of DWV into the body of honey bee larvae enabled amplification of the virus and a rapid emergence of DWV virulent strains [DWV-VDV-1-like recombinants (Gisder et al. 2009; Ryabov et al. 2014)].

These data enable us to hypothesize that the mite contribution to the emergence of virulent strains of DWV could be:

1. The rapid accumulation of a variety of DWV variant strains that may even replicate in the body of the mite and,
2. Their subsequent injection directly into the bee hemolymph, overcoming the primary immune defenses of the bee which are normally directed toward pathogens naturally introduced by oral ingestion, a route known to be much less effective (Mockel et al. 2011).

A summary of the Varroa-DWV vicious cycle is presented in Fig. 1.

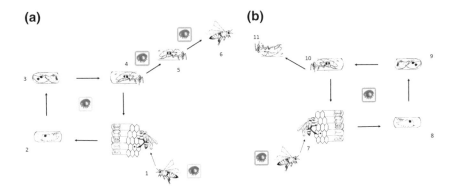

Fig. 1 The Varroa–DWV cycle. **a** Infestation of a new colony by Varroa carrying low DWV loads (blue Varroa) and conversion to highly virulent DWV. 1. Brood infestation. 2. Reproduction. 3. Reproduction and amplification of the viral load. 4. Amplification of DWV-virus injected to the bee hemolymph induces appearance of highly virulent DWV. 5. Emerging bees carrying Varroa with highly virulent DWV (red Varroa) can transmit the DWV-loaded parasite and/or DWV to other bees from the same colony. 6 Foraging bees can transmit the highly DWV-loaded Varroa to other colonies. **b** The Varroa–DWV cycle re-initiates with the red Varroa in the same or in other colonies and brings them to their collapse

5   CBPV Opportunistic Infections and Mechanical Stress

A common observation is that CBPV infections erupt often when the colony seems to be strong (a robust population of adults and brood). Frequently, this eruption was attributed to the mechanical break of the bees' body hairs due to overcrowding of the colony population before swarming; such breakage could facilitate the access of contaminating CBPV to the bee hemolymph (Ribiere et al. 2010).

Another example of stress-induced infections was observed after beehives were put under nets and into greenhouses for pollination. Keeping up with the increasing trend in Israel to utilize honey bee hives for pollination in net-covered crops, we noticed an increase in piles of dead bees in the front of those hives. These bees displayed the characteristics typical of CBPV-induced paralysis and death. Diagnosis performed in the laboratory showed that they were highly infected with CBPV (viral titers of above 10^9 particles per bee). To confirm our initial findings, we introduced a group of colonies at the entrance of net-covered crops at two locations in the country, and kept an equal number of control colonies uncovered, at open crop conditions. We found that the hives located at the covered crops' entrance quickly contracted CBPV (Slabezki, Y, Dag, A. and Chejanovsky N, Manuscript in preparation).

These findings support the hypothesis that mechanical stress caused cuticular damage to the pollen/nectar-loaded honey bee foragers by their collision with the nets in an attempt to return to the hive, providing the virus quick access to the insect hemolymph, and thus overcoming the insect defenses.

6   DWV and Insecticide Exposure—Insecticide Spread and Virus Emergence

Some insecticides were documented as causing stress responses in honey bees (Blacquiere et al. 2012). This resulted in temporarily banning the use of three neonicotinoids by the European Union (Gross 2013). A recent study showed that application of the neonicotinoid clothianidin weakened the immune defenses of recently emerged worker bees (Di Prisco et al. 2013). Furthermore, it involved the repression of expression of another member of the NfκB family (Di Prisco et al. 2013). Under these circumstances, DWV-dormant infections with low levels of viral replication were promoted to replicate DWV at high levels, comparable to those observed in symptomatic infections.

The stress situations presented above referred to induction of particular viruses. However, we and others have observed the simultaneous or progressive appearance of several pathogenic viruses upon weakening of honey bee defenses by biological, chemical, or environmental stresses. These superinfections then contribute to the rapid deterioration of the colony.

7   Prophylaxis Methods and Antiviral Approaches

7.1   What Can We Learn from Stress-Induced Infections?

The three cases discussed in detail above exposed a link between stress induction, weakening of the immune system of the bees, and the activation of lethal viral infections (summarized in Fig. 2) and suggest that if we adopt appropriate measures, we should be able to maintain the damage to colonies at sustainable levels. If we "beekeepers" look at treatment according to the different elements that can co-act to weaken a colony, beekeepers should be able to attain a comprehensive treatment.

7.2   Can We Treat Viral Infections?

From the point of view of virus treatments, we should aim to reduce:

1. The conversion of avirulent strains to virulent strains.
2. Block the replication of viruses.
3. Reduce the possibility of their transmission.

7.2.1   Conversion of Avirulent Strains to Virulent Strains

As we discussed above, Varroa is an active vector of viruses and promotes their direct access to the host hemolymph overcoming bee defenses. This direct infection

Fig. 2 Biological (e.g. mites), chemical (insecticides, miticides), and mechanical stressors can induce dormant and new honey bee virus infections

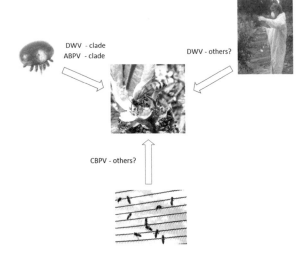

route facilitates conversion of viruses from avirulent to virulent strains. To interrupt this process, we (beekeepers) should aim to diminish the effect of Varroa by controlling the mites. The timing and dose of treatment should be applied early in the season, which is crucial to avoid virus conversion (from low virulent to highly virulent strains).

However, some Varroa treatments can induce stress in bees and incidentally increase the virus too. For example, it has been reported that coumaphos and fluvalinate treatments can induce changes in certain honey bee genes related to immunity, detoxification, behavioral maturation, and nutrition (Boncristiani et al. 2012; Schmehl et al. 2014). Thus, applying thoughtful, professionally assisted treatments against Varroa (and even alternating control measures to avoid the emergence of resistant Varroa mites) could diminish its long-term impact on bee health. Successful breeding of bees to resist Varroa infestation might achieve similar results (Locke et al. 2014; Rinderer et al. 2010; Buchler et al. 2010).

Knowledge of the insecticides used on the crops in the vicinity to the bee colonies could prevent the replication of undesirable highly pathogenic viruses. From a long-term perspective, it will be important to coordinate insecticide applications (type and timing) with honey bee colony placement and more strongly advocating the use of honey bee-friendly insecticides (if these even exist).

Proper nutrition, such as pollen (or protein-based feeding), was shown to reduce the impact of the pesticide chlorpyrifos; and it can help to reduce the negative impact of some insecticides (Schmehl et al. 2014). In contrast, excessive reliance on feeding sugar syrup may have detrimental effects, since they may have a negative impact on the performance of the honey bee immune system (Galbraith et al. 2015).

In other cases, such as mechanical stress, using nets that could be less damaging to the bees, or even working in other types of covered crops, would help.

7.2.2 Block Replication of Viruses

Viruses replicate only in the body of their hosts, yielding high numbers of viral particles (virions) that propagate the infection. During this process, multiple copies of the viral genome that encodes the genetic information for its propagation are produced (from a million to a billion copies). In addition, new variations of the original information (variants) are produced, and the chance that these variants convert from an avirulent to a more virulent virus and infectious strains (viral strain) increases with the increase in the number of virions produced.

Thus, blocking the ability of viruses to replicate may reduce the chance of the emergence of more virulent and infectious strains. Sometimes the host (in this case the honey bee) can develop such blocking, but little is known about the ability of different bee races to resist viral infections. Much more research needs to be supported in these areas.

On the other hand, a promising approach is based on the fact that it is possible to target the replication of honey bee viruses by utilizing biological tools that mimic or enhance the host immune response. This approach is based on what is known as the RNA interference response (Niu et al. 2014). This natural response detects the presence of foreign (non-host) RNA, such as the genome of RNA viruses of the honey bee, and promotes their specific degradation when the virus is trying to replicate (Niu et al. 2014). During this process, the viral genome is chopped into useless pieces by the honey bee immune defense mechanism. It became clear that it is possible to induce this response by producing in vitro (in the laboratory) molecules of double-strand RNA. Double-stranded RNAs, or (dsRNAs), are short molecules with one strand and its mirror copy). Such RNA strands carry small bits of the genetic information for specific viruses. Subsequent injections or feeding of these RNA molecules to honey bees triggered the RNAi response. This resulted in the inhibition of the ability of the virus to replicate in the honey bee. This was shown for SBV, DWV, and IAPV (Desai et al. 2012; Liu et al. 2010; Maori et al. 2009). Furthermore, in the case of IAPV, it was revealed that the administration of dsRNA protected the colonies from viral infection (Hunter et al. 2010). However, in the latter case, RNAi was administrated concurrently with IAPV, and it remains to be demonstrated that its application postinfection is efficient to diminish/ameliorate viral damage (Hunter et al. 2010).

From the very beginning of animal and plant virus research in the middle of the last century, viruses were considered as mysterious pathogens. However, research has produced drugs and treatments against a series of serious viral pathogens, such as the flu viruses, herpes viruses, human immunodeficiency viruses, small pox virus. These treatments were aimed at stopping the multiplication of the viruses. Thus, it is conceivable that in the future, there will be progress in understanding the replication of honey bee viruses which may yield experimental drugs that could block virus infections or immunize honey bees.

7.2.3 Reduce Their Transmission

Early monitoring of symptomatic viral infections can be used as a preventative measurement. In the case of DWV, for example, observant beekeepers could recognize that the more virulent viral strains were already present (at least in some colonies), and measures can be taken to prevent their spread.

8 Conclusions

There is much more to be learned about bee viruses, such as the natural resistance of different honey bee strains against virus infections, the genetic basis of this resistance, and the effect of human-borne and environmental factors that can upset or maintain bee virus infections. Current research is trying to understand more about these processes.

In the meantime, we hope that this material presented an overall view of virus infections associated with colony losses, the stress factors involved in their acute manifestation, and possible measures that can contribute to ameliorating their impact on your honey bee operation.

Acknowledgements We gratefully acknowledge Reut Bernheim for her support in drawing Fig. 1. This work was partially supported by Grants of the Chief Scientist of the Ministry of Agriculture (NC, YS) number 131-1723 and 131-1815.

Glossary

Genome A DNA or RNA molecule, depending on the virus, bearing all the information the virus needs to replicate in the cells of its host
Viral genomic copies Number of viral genomes that bear the genetic information that allows the virus to produce more viral particles
Viral loads Usually refers to the number of viral genomic copies which is the most common method of estimating honey bee viruses, but it could also refer as well to the number of infectious virus particles
Viral genomic replication The process by which the virus produces new copies, replicas, of itself, that are packed in new viral particles
Immunosuppression Weakening of the immune system, body defenses
Down-regulation of genes A molecular process that results in lower expression of the proteins that are products of these genes
Genomic homology Similarity of nucleotide sequences between virus genomes

References

Amiri E, Meixner M, Buchler R, Kryger P (2014) Chronic bee paralysis virus in honeybee queens: evaluating susceptibility and infection routes. Viruses-Basel 6(3):1188–1201. doi:10.3390/v6031188

Bailey L, Woods RD, Gibbs AJ (1963) Two viruses from adult honey bees (*Apis mellifera* Linnaeus). Virology 21:390–395

Bailey L, Ball BV, Perry JN (1981) The prevalence of honey bee viruses in Britain. Ann Appl Biol 97:109–118

Berthoud H, Imdorf A, Haueter M, Radloff S, Neumann P (2013) Virus infections and winter losses of honey bee colonies (*Apis mellifera*). J Apic Res 49:60–65

Blacquiere T, Smagghe G, van Gestel CAM, Mommaerts V (2012) Neonicotinoids in bees: a review on concentrations, side-effects and risk assessment. Ecotoxicol 21:973–992

Blanchard P, Ribiere M, Celle O, Lallemand P, Schurr F, Olivier V, Iscache AL, Faucon JP (2007) Evaluation of a real-time two-step RT-PCR assay for quantitation of Chronic bee paralysis virus (CBPV) genome in experimentally-infected bee tissues and in life stages of a symptomatic colony. J Virol Methods 141:7–13

Boecking O, Genersch E (2008) Varroosis—The ongoing crisis in bee keeping. J Verbraucher-schutz und Lebensmittelsicherheit 3:221–228

Boncristiani HF, di Prisco G, Pettis JS, Hamilton M, Chen YP (2009) Molecular approaches to the analysis of deformed wing virus replication and pathogenesis in the honey bee, *Apis mellifera*. Virology J 6:221. doi:10.1186/1743-422X-6-221

Boncristiani H, Underwood R, Schwarz R, Evans JD, Pettis J, Vanengelsdorp D (2012) Direct effect of acaricides on pathogen loads and gene expression levels in honey bees *Apis mellifera*. J Insect Physiol 58:613–620

Boncristiani HF, Evans JD, Chen Y et al (2013) In vitro infection of pupae with Israeli acute paralysis virus suggests disturbance of transcriptional homeostasis in honey bees (*Apis mellifera*). PLoS ONE 89:e73429. doi:10.1371

Buchler R, Berg S, le Conte Y (2010) Breeding for resistance to *Varroa destructor* in Europe. Apidologie 41:393–408. doi:10.1051/apido/2010011

Chen YP, Siede R (2007) Honey bee viruses. Adv. Virus. Res. 70:33–80

Chen YP, Pettis JS, Corona M, Chen WP, Li CJ, Spivak M, Visscher PK, Degrandi-Hoffman G, Boncristiani H, Zhao Y, Vanengelsdorp D, Delaplane K, Solter L, Drummond F, Kramer M, Lipkin WI, Palacios G, Hamilton MC, Smith B, Huang SK, Zheng HQ, Li JL, Zhang X, Zhou AF, Wu LY, Zhou JZ, Lee ML, Teixeira EW, Li ZG, Evans JD (2014) Israeli acute paralysis virus: epidemiology, pathogenesis and implications for honey bee health. PLoS Pathog 10(7):e1004261. doi:10.1371/journal.ppat.1004261

Cornman RS, Tarpy DR, Chen Y, Jeffreys L, Lopez D, Pettis JS, van Engelsdorp D, Evans JD (2012) Pathogen webs in collapsing honey bee colonies. PLoS ONE 7(8):e43562. doi:10.1371/journal.pone.0043562

Cox-Foster DL, Conlan S, Holmes EC, Palacios G, Evans JD, Moran NA, Quan PL, Briese T, Hornig M, Geiser DM, Martinson V, Vanengelsdorp D, Kalkstein AL, Drysdale A, Hui J, Zhai J, Cui L, Hutchison SK, Simons JF, Egholm M, Pettis JS, Lipkin WI (2007) A metagenomic survey of microbes in honey bee colony collapse disorder. Science 318:283–287

Dainat B, Evans JD, Chen YP, Gauthier L, Neumann P (2012a) dead or alive: deformed wing virus and *Varroa destructor* reduce the life span of winter honeybees. Appl Environ Microbiol 78 (4):981–987. doi:10.1128/AEM.06537-11

Dainat B, Evans JD, Chen YP, Gauthier L, Neumann P (2012b) Predictive markers of honey bee colony collapse. PLoS ONE 7(2):e32151. doi:10.1371/journal.pone.0032151

de Miranda JR, Genersch E (2010) Deformed wing virus. J Invertebr Pathol 103(Suppl. 1):S48–S61. doi:10.1016/j.jip.2009.06.012

de Miranda JR, Cordoni G, Budge G (2010) The acute bee paralysis virus-Kashmir bee virus-Israeli acute paralysis virus complex. J Invertebr. Pathol 103(Suppl. 1):S30–S47. doi:10.1016/j.jip.2009.06.014

de Miranda JR, Bailey L, Ball BV et al (2013) Standard methods for virus research in *Apis mellifera*. J Apic Res 52(4). doi:52.4.22

Desai SD, Eu YJ, Whyard S, Currie RW (2012) Reduction in deformed wing virus infection in larval and adult honey bees (*Apis mellifera* L.) by double-stranded RNA ingestion. Insect Mol Biol 21(4):446–455. doi:10.1111/j.1365-2583.2012.01150.x

di Prisco G, Cavaliere V, Annoscia D, Varricchio P, Caprio E, Nazzi F, Gargiulo G, Pennacchio F (2013) Neonicotinoid clothianidin adversely affects insect immunity and promotes replication of a viral pathogen in honey bees. Proc Natl Acad Sci USA 110(46):18466–18471. doi:10.1073/pnas.1314923110

Fievet J, Tentcheva D, Gauthier L, de Miranda J, Cousserans F, Colin ME, Bergoin M (2006) Localization of deformed wing virus infection in queen and drone *Apis mellifera* L. Virology J 3:16. doi:10.1186/1743-422X-3-16

Fujiyuki T, Takeuchi H, Ono M et al (2005) Kakugo virus from brains of aggressive worker honeybees. In: Maramorosch K, Shatkin AJ (eds) Adv Virus Res Vol 65

Galbraith DA, Yang X, Nino EL, Yi S, Grozinger C (2015) Parallel epigenomic and transcriptomic responses to viral infection in honey bees (*Apis mellifera*). PLoS Pathog 11(3):e1004713. doi:10.1371/journal.ppat.1004713

Gauthier L, Tentcheva D, Tournaire M, Dainat B, Cousserans F, Colin ME, Bergoin M (2007) Viral load estimation in asymptomatic honey bee colonies using the quantitative RT-PCR technique. Apidologie 38:426–435

Genersch E, von der Ohe W, Kaatz H, Schroeder A, Otten C, Buchler R, Berg S, Ritter W, Muhlen W, Gisder S, Meixner M, Liebig G, Rosenkranz P (2010) The German bee monitoring project: a long term study to understand periodically high winter losses of honey bee colonies. Apidologie 41:332–352

Gisder S, Aumeier P, Genersch E (2009) Deformed wing virus: replication and viral load in mites (*Varroa destructor*). J Gen Virol 90(Pt 2):463–467. doi:10.1099/vir.0.005579-0

Gross M (2013) EU ban puts spotlight on complex effects of neonicotinoids. Curr Biol 23(11): R462–R464. doi:10.1016/j.cub.2013.05.030

Highfield AC, el Nagar A, Mackinder LCM, Noel L, Hall MJ, Martin SJ, Schroeder DC (2009) Deformed wing virus implicated in overwintering honeybee colony losses. Appl Environ Microbiol 75(22):7212–7220. doi:10.1128/AEM.02227-09

Hou CS, Rivkin H, Slabezki Y, Chejanovsky N (2014) Dynamics of the presence of Israeli acute paralysis virus in honey bee colonies with colony collapse disorder. Viruses-Basel 6(5):2012–2027. doi:10.3390/v6052012

Hunter W, Ellis J, Vanengelsdorp D, Hayes J, Westervelt D, Glick E, Williams M, Sela I, Maori E, Pettis J, Cox-Foster D, Paldi N (2010) Large-scale field application of RNAi technology reducing Israeli acute paralysis virus disease in honey bees (*Apis mellifera*, Hymenoptera: Apidae). PLoS Pathog 6(12):e1001160. doi:10.1371/journal.ppat.1001160

Iqbal J, Mueller U (2007) Virus infection causes specific learning deficits in honeybee foragers. Proc R Soc London B 274:1517–1521

Li ZG, Chen YP, Zhang SW, Chen SL, Li WF, Yan LM, Shi LG, Wu LM, Sohr A, Su SK (2013) Viral infection affects sucrose responsiveness and homing ability of forager honey bees, *Apis mellifera* L. Plos One 8(10):e77354. doi:10.1371/journal.pone.0077354

Liu XJ, Zhang Y, Yan X, Han RC (2010) Prevention of Chinese sacbrood virus infection in *Apis cerana* using RNA interference. Curr Microbiol 61:422–428

Locke B, Forsgren E, de Miranda JR (2014) Increased tolerance and resistance to virus infections: a possible factor in the survival of *Varroa destructor*-resistant honey bees (*Apis mellifera*). PLoS ONE 9(6):e99998. doi:10.1371/journal.pone.0099998

Maori E, Lavi S, Mozes-Koch R, Gantman Y, Peretz Y, Edelbaum O, Tanne E, Sela I (2007) Isolation and characterization of Israeli acute paralysis virus, a dicistrovirus affecting honeybees in Israel: evidence for diversity due to intra- and inter-species recombination. J Gen Virol 88:3428–3438

Maori E, Paldi N, Shafir S, Kalev H, Tsur E, Glick E, Sela I (2009) IAPV, a bee-affecting virus associated with colony collapse disorder can be silenced by dsRNA ingestion. Insect Mol Biol 18(1):55–60. doi:10.1111/j.1365-2583.2009.00847.x

Martin SJ, Highfield AC, Brettell L, Villalobos EM, Budge GE, Powell M, Nikaido S, Schroeder DC (2012) Global honey bee viral landscape altered by a parasitic mite. Science (New York, N.Y.) 336:1304–1306

Martin SJ (2001) The role of Varroa and viral pathogens in the collapse of honeybee colonies: a modelling approach. J Appl Ecol 38 (5):1082–1093

Mockel N, Gisder S, Genersch E (2011) Horizontal transmission of deformed wing virus: pathological consequences in adult bees (Apis mellifera) depend on the transmission route. J Gen Virol 92(Pt 2):370–377. doi:10.1099/vir.0.025940-0

Moore J, Jironkin A, Chandler D, Burroughs N, Evans DJ, Ryabov EV (2011) Recombinants between deformed wing virus and Varroa destructor virus-1 may prevail in Varroa destructor-infested honeybee colonies. J Gen Virol 92(Pt 1):156–161. doi:10.1099/vir.0.025965-0

Nazzi F, Brown SP, Annoscia D et al (2012) Synergistic parasite-pathogen interactions mediated by host immunity can drive the collapse of honeybee colonies. PLoS pathogens 8(6):e1002735. doi:10.1371/journal.ppat.1002735

Niu J, Meeus I, Cappelle K, Piot N, Smaghe G (2014) The immune response of the small interfering RNA pathway in the defense against bee viruses. Curr Opin Insect Sci 6:22–27. doi:10.1016/j.cois.2014.09.014

Ongus JR, Peters D, Bonmatin JM, Bengsch E, Vlak JM, van Oers MM (2004) Complete sequence of a picorna-like virus of the genus Iflavirus replicating in the mite Varroa destructor. J Gen Virol 85:3747–3755

Ribiere M, Lallemand P, Iscache AL, Schurr F, Celle O, Blanchard P, Olivier V, Faucon JP (2007) Spread of infectious chronic bee paralysis virus by honeybee (Apis mellifera L.) feces. Appl Environ Microbiol 73:7711–7716

Ribiere M, Olivier V, Blanchard P (2010) Chronic bee paralysis: A disease and a virus like no other? J Invertebr Pathol Suppl. 103(1):S120–S131. doi:10.1016/j.jip.2009.06.013

Rinderer TE, Harris JW, Hunt GJ, de Guzman LI (2010) Breeding for resistance to Varroa destructor in North America. Apidologie 41:409–424

Runckel C, Flenniken ML, Engel JC, Ruby JG, Ganem D, Andino R, Derisi JL (2011) Temporal analysis of the honey bee microbiome reveals four novel viruses and seasonal prevalence of known viruses, Nosema, and Crithidia. PLoS ONE 6(6):e20656. doi:10.1371/journal.pone.0020656

Ryabov EV, Wood GR, Fannon JM, Moore JD, Bull JC, Chandler D, Mead A, Burroughs N, Evans DJ (2014) A virulent strain of deformed wing virus (DWV) of honeybees (Apis mellifera) prevails after Varroa destructor-mediated, or in vitro transmission. Plos Pathogens 10(6):e1004230. doi:10.1371/journal.ppat.1004230

Schmehl DR, Teal PEA, Frazier JL, Grozinger CM (2014) Genomic analysis of the interaction between pesticide exposure and nutrition in honey bees (Apis mellifera). J Insect Physiol 71:177–190. doi:10.1016/j.jinsphys.2014.10.002

Shah KS, Evans EC, Pizzorno MC (2009) Localization of deformed wing virus (DWV) in the brains of the honeybee, Apis mellifera Linnaeus. Virology J 6:182. doi:10.1186/1743-422X-6-182

Shen M, Yang X, Cox-Foster D, Cui L (2005) The role of varroa mites in infections of Kashmir bee virus (KBV) and deformed wing virus (DWV) in honey bees. Virology 342:141–149

Soroker V, Hetzroni A, Yakobson B, David D, David A, Voet H, Slabezki Y, Efrat H, Levski S, Kamer Y, Klinberg E, Zioni N, Inbar S, Chejanovsky N (2011) Evaluation of colony losses in Israel in relation to the incidence of pathogens and pests. Apidologie 42:192–199

Stokstad E (2007) The case of the empty hives. Science 316:970–972

Tentcheva D, Gauthier L, Bagny L, Fievet J, Dainat B, Cousserans F, Colin ME, Bergoin M (2006) Comparative analysis of deformed wing virus (DWV) RNA in Apis mellifera and Varroa destructor. Apidologie 37:41–50

Toplak I, Ciglenecki UJ, Aronstein K et al (2013) Chronic bee paralysis virus and Nosema ceranae experimental co-infection of winter honey bee workers (Apis mellifera L.). Viruses-Basel, 5(9), 2282–2297. doi:10.3390/v5092282

van Engelsdorp D, Evans JD, Saegerman C, Mullin C, Haubruge E, Nguyen BK, Frazier M, Frazier J, Cox-Foster D, Chen YP, Underwood R, Tarpy DR, Pettis JS (2009) Colony Collapse disorder: a descriptive study. PLoS ONE 4(8):e6481. doi:10.1371/journal.pone.0006481
Yue C, Gensersch E (2005) RT-PCR analysis of deformed wing virus in honeybees (*Apis mellifera*) and mites (*Varroa destructor*). J Gen Virol 86:3419–3424
Zioni N, Soroker V, Chejanovsky N (2011) Replication of Varroa destructor virus 1 (VDV-1) and a Varroa destructor virus 1-deformed wing virus recombinant (VDV-1-DWV) in the head of the honey bee. Virology 417(1):106–112. doi:10.1016/j.virol.2011.05.009

Author Biographies

Nor Chejanovsky is an insect virologist. He obtained his Ph.D. at The Hebrew University of Jerusalem, Israel. He performed his postdoctoral studies at the National Institutes of Health Maryland, USA. In 2005, he spent 1-year sabbatical at Harvard Medical School in Boston, USA. Actually, he is on sabbatical at the Institute of Bee Health at the Vetsuisse, University of Bern, Switzerland. He is a member of COLOSS (the international network on Colony Losses) where he is leading the Virus Task Force with colleagues. His laboratory at the Department of Entomology in The Volcani Center, Israel, is dedicated to the study of insect viruses since 1989 and specifically honey bee viruses since 2007. He focuses on virus-honey bee and virus-Varroa relationships utilizing molecular tools, laboratory bioassays, and field data, aiming to ameliorate virus damage to honey bee colonies. He teaches Virology and Insect Pathology at the Faculty of Agriculture of The Hebrew University of Jerusalem. He closely collaborates with the Israeli Ministry of Agriculture Extension Service, beekeepers, and researchers worldwide.

Yossi Slabezki finished his studies at The Faculty of Agriculture from The Hebrew University of Jerusalem in 1987, specializing in entomology and beekeeping. His research thesis was on factors affecting honey bee swarming. Since then he works at the Israeli Ministry of Agriculture Extension Service in the beekeeping branch focused on resolving honey beekeeping problems: acarology, virology, toxicology, etc. His endeavor is to link between the field and research front back and forth. From this interchange emerge and are implemented the treatments' policy of the honey bee branch in Israel. During these years, Yossi specialized in honey bee pathogens and participated in research projects on *Varroa destructor*, tracheal mites, honey bee viruses, and Nosema. He is also actively involved in prevention of insecticide damage and intoxication of bees as well as in efforts to enlarge the bee-safe foraging areas. Actually, he is the Director of the Apiculture Department at the Israeli Ministry of Agriculture Extension Service and Lecturer of beekeeping at The Faculty of Agriculture of The Hebrew University of Jerusalem.

Using Epidemiological Methods to Improve Honey Bee Colony Health

Nathalie Steinhauer and Dennis vanEngelsdorp

Abstract

Epidemiologists emphasize that health is a common good. By focusing research on health at the population level, epidemiology can make great positive impact on health. Using real-world examples, we hope to give you a quick overview of what epidemiology is, how it works and should be interpreted. First off, epidemiology is all about measuring (1) how much disease there is and (2) what factors contribute to the occurrence or absence of disease. So if you are a beekeeper, and you want to keep your bees alive (and why wouldn't you?), you should first understand the ways disease and risk are calculated and used to develop strategies to maximize bee health. This chapter is meant to do just that—gives a quick primer of epidemiology—so you and your fellow beekeepers have some way of self-evaluating new and old research about bee health and management and figure out how to apply new knowledge when managing your colonies to maximize their health.

Honey bees have been dying in the USA and in other countries at high rates for over a decade (Laurent et al. 2015; Seitz et al. 2015). Beekeepers have questions: *How many managed honey bee colonies died last winter in the USA? How did my operation compare? How many Nosema spores per bee are needed to justify treatment? What number of Varroa mites can live in my colony without hurting it? How should I treat an outbreak of European foulbrood? What can I do to reduce the chances of getting American foulbrood?* These are the kinds of questions epidemiologists try to answer.

N. Steinhauer (✉) · D. vanEngelsdorp
Department of Entomology, University of Maryland, 4112 Plant Science Building,
College Park, MD 20742, USA
e-mail: nsteinha@umd.edu

© Springer International Publishing AG 2017
R.H. Vreeland and D. Sammataro (eds.), *Beekeeping – From Science to Practice*,
DOI 10.1007/978-3-319-60637-8_8

Using real-world examples, we hope to give you a quick overview of what epidemiology is and how it works. First off, epidemiology is all about measuring (1) how much disease there is and (2) what factors contribute to the occurrence or absence of disease. So if you are a beekeeper, and you want to keep your bees alive (and why wouldn't you?), you should first understand the ways disease and risk are calculated and used to develop strategies to maximize bee health. This chapter is meant to do just that—gives a quick primer of epidemiology—so you and your fellow beekeepers have some way of self-evaluating new and old research about bee health and management and figure out how to apply new knowledge when managing your colonies.

1 What Is Epidemiology?

Epidemiology is the study of disease levels in a population. Epidemiologists use a broad definition of "disease": any departure from perfect health. Honey bees pose a particular problem for epidemiologists as it is hard to define what a colony in perfect health would look like. Fortunately, diseased colonies are easier to identify.

When measuring disease or a departure from perfect health—we use both direct and indirect measures. **Direct** measures are easiest to understand. You go to a hive and see symptoms that look like American foulbrood (AFB), you take samples and send them to a laboratory, and the samples come back positive: Your colony has AFB. But as all beekeepers know, honey bee colonies are complicated, and since we can't ask the bees how they are feeling we have to use several **indirect** measurements to assess how healthy a colony is. Good beekeepers do this every time they inspect colonies—what does the brood pattern look like? How many frames of bees and brood and food are present? How is the queen doing? Is the colony alive? Obviously, a dead colony is the worst and most extreme "disease" outcome there is.

To be honest, epidemiologists are not really interested in how disease might affect a single individual (another epidemiological challenge—What is an individual in the beekeeping context? A bee? A colony? An apiary?—but more on that later), rather, epidemiologists focus on how disease spread and persist (or not!) in a population. The ultimate goal of epidemiologist is disease prevention, and so epidemiologists also evaluate prevention strategies and devise and evaluate ways to get the proven best practices widely adopted.

At the core, there are two different types of epidemiological studies, descriptive studies and analytic studies. As their name suggests, **descriptive studies** are designed to describe a disease, how widespread it is, where it occurs, when it occurs, etc. These studies do not necessarily try to link disease outcomes with cause (s). **Analytical studies,** on the other hand, are designed to determine which factors are related to disease outcomes. By measuring "exposure variables" (also called, "risk factors") and the occurrence (or not!) of a disease, analytical studies quantify the chances an individual will develop a particular disease after a certain exposure.

So if we think again about AFB, a descriptive study would endeavor to find out how many diseased colonies there are in a certain area over a certain period of time, while an analytical study would attempt to find factors that increased the chances that a colony would have AFB. Sometimes these risk factors are self-evident, e.g., buying used equipment from a neighbor who had a major AFB problem and did not know it.

The key point of analytical studies is to measure the **risk** or **chance** of developing disease after an exposure. It is important to remember that in many cases not all colonies exposed to a risk factor will get the disease, and not all colonies suffering from a disease will have been exposed to the same risk factor(s). Using our AFB example again, after being exposed to AFB spore-contaminated brood comb (Lindström et al. 2008a), two of five colonies developed AFB infections, meaning, three colonies on five did not show clinical symptoms of AFB even when sharing the same exposure. Conversely, not all cases of AFB are the result of introducing contaminated comb into uninfected colonies, for example bees can rob honey from infected colonies and bring the pathogen home starting an infection (Lindström et al. 2008b).

a. **Surveillance and monitoring in honey bee health**

Surveying a population for disease is the most basic form of a descriptive study. Systematic surveys conducted over time can help define "normal" disease rates in a population. Importantly, once we know what normal disease levels are, survey results help identify outbreaks and/or hotspots of disease occurrence. As one can imagine, finding a new disease early, before it spreads widely, is the reason many surveillance efforts are implemented. Knowledge of where and when a disease emerges is the starting point for many epidemiological investigations.

Beekeepers do this informally all the time. Every time you open and inspect a colony and look for evidence of brood disease you are "surveying" your operation. Of course other surveillance efforts, such as conducted by state apiary inspectors, are more structured and systematically look for disease within the operations of their purview on a regular and long-term basis. For survey data to have the most value, clear protocols are required so that data from many different inspections are comparable. Such data, aggregated over time and space, has huge value, as it can compare bee health over time and also provide insight on the relationship between colony health measures.

Illustration 1. **Apiary Inspections: An Example of Surveillance Program**. In the early 1900s, in response to the high prevalence of the highly contagious "foul brood" (this was before the bacteria responsible was identified and the condition re-named "American Foulbrood") in the USA, many USA states enacted bee laws that mandated the inspection of honey bee colonies on a regular basis to help find and then destroy colonies that were contaminated (Burgess and Howard 1906). The Pennsylvania Department of Ag kept records of AFB prevalence that date back to the 1940s. When the survey was first conducted, AFB was found in 12% of all the apiaries inspected. In the 2000s, it was well below 1% (unpublished). Note that this data says nothing about why the disease rate went down.

Systematic monitoring allows for early detection—and reaction—to the appearance of newly emerging diseases. Let's be honest, bees have faced and continue to face a lot of threats. Over 100 years ago, one of the biggest problems faced by North American beekeepers was Wax moth. Since then, we have had to face AFB, chalkbrood, sac brood, honey bee tracheal mite, *Varroa*, small hive beetle, more than a dozen virus; and we are now threatened by Asian hornets (racking havoc in France and other places in Europe), even more viruses, and the Tropilaelaps mite. The accidental introduction of Tropilaelaps mite into any country is the one threat that should keep beekeepers awake at night. Like *Varroa*, it evolved on a different species of honey bee and jumped host. In some places in Asia where they keep European honey bees, they have to treat for the mite once every 2 weeks (Pettis et al. 2013)! The value of a surveillance system (reviewed in Lee et al. 2015a) is to ensure that if Tropilaelaps mites are introduced into the USA, detecting it early would allow for interventions which would hopefully eradicate the problem before it became wide spread.

Box 1: Measures of disease frequency
Epidemiologists have their own jargon, which you will likely encounter in study summaries or reports. Here are some of the most common and useful terms defined.

 Prevalence and *incidence* are two measures of disease in a population, and so usually the main result of descriptive studies (see Fig. 1). **Prevalence** is the proportion (usually expressed as percentage) of existing cases in a known population. It indicates how frequently a disease is present in a population during the survey time frame. If the survey randomly selected colonies to inspect, one way to interpret prevalence is the probability that any subject in a population has the disease. **Incidence** is the number of new cases that developed over a specified period of time in the population at risk. It specifically relates to the transition from a healthy state to a diseased state rather than just the number of diseased individuals. In the apiary depicted in Fig. 1, the prevalence of the "disease" was 37.5% (three of eight) on Date 1. The second inspection found that four of the seven were infected, so the prevalence was 57.1%. During the interval between Date 1 and Date 2, the incidence of the disease was 60% (three new infections in the five that were "at risk"—those that were not infected during the first inspection). Mortality rates are also a form of incidence. In the case we have just discussed, all eight colonies originally inspected were at risk of dying, one did die so the incidence (or mortality) rate was 12.5% (one of eight) for the period of time between Date 1 and Date 2.

Apiary

Fig. 1 Incidence and prevalence. Fictitious apiary represented at two different dates. *Legend Green* colonies = disease absent; *Red* colonies = disease present; *Black* colonies = colony died. One Date 1, three of the eight colonies are diseased. The prevalence of the disease is 3/8 = 37.5% on Date 1. On Date 2, one of the diseased colonies has recovered, one is lost and three previously healthy colonies became diseased. The prevalence of the disease on Date 2 is 4/7 = 57.1%. The incidence of the disease between Date 1 and Date 2 is of 3/5 = 60% (three new cases among the five healthy, at risk, colonies on Date 1)

Disease Loads

Disease frequency (see Box 1) is useful in understanding bee health in some context but not others. When a disease is ubiquitous, then not much useful information is gained by just knowing if colonies have or do not have the disease. In other words, sometimes it is not the prevalence that matters, it is the disease's **load** that matters. For most diseases, and in particular infectious disease, the gravity of infection in diseased individuals provides a more complete picture than the simple presence or absence of the disease.

For example, nearly every colony in the continental USA has *Varroa*, specifically between 2011 and 2015 *Varroa* were detected in 91.7% of colonies sampled (i.e., 91.7% prevalence, Traynor et al. 2016). Please note we specifically and deliberately used the word *detected*—as, in all likelihood, the prevalence was much greater and the negative detections probably were the result of recent *Varroa* treatment applied by the beekeeper before inspection, so that, while mites were present they were at very low—undetectable—levels. We discuss the idea of test sensitivity and specificity in Box 2. If about every colony is infected with *Varroa*, then there is little value in knowing the prevalence (mere presence/absence) in order to help understand bee health. The real data of import is how many *Varroa* are found in a colony—the colonies mite **load** (Fig. 2). This information is predictive and actionable.

Values related to disease loads may be used as the trigger for management decisions (such as deciding which colonies to use as breeder stock, or when to apply a chemical or non-chemical control strategy). The *Varroa* load that currently warrants action—the **action threshold**—is a little tricky to pin down. It depends—on the time of the year (high levels early in the season are more concerning then a comparable level latter in the season), on the viruses the mites are vectoring

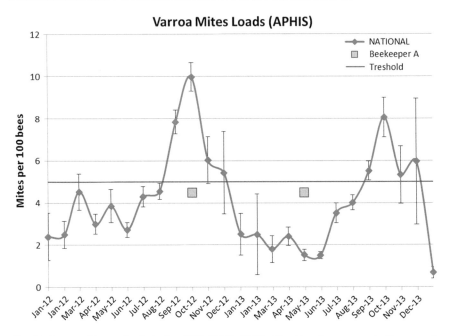

Fig. 2 Average *Varroa* loads in the USA (2012–2013). The *blue line* represents *Varroa* mite loads observed from the *USDA APHIS National Honey Bee Disease Survey* 2012–2013 ($n = 1515$ operations sampled; data from Traynor et al. 2016), averaged by month of observation (with error bars as standard errors). The two *orange squares* represent fictitious samples sent in by a beekeeper (used in the discussion above)

(*Varroa* can spread very fast which can wipe out much of an operation even when mite levels remain low), and on the region (with or without an interruption of brood due to winter). Disease loads often have a seasonality (Fig. 2). Knowledge of this seasonality is helpful when designing management tools to reduce losses. Mite levels peak in the USA population in late summer/early fall. This is also the time that colonies begin to crash from heavy mite infestations. When a colony crashes from *Varroa*, many of those mites are spread to neighboring colonies by the drift of the collapsing colonies last bees or by robbing bees that pick up *Varroa* while plundering the honey reserves of the collapsing colony (Frey and Rosenkranz 2014). For this reason, beekeepers are urged to check mite levels in hives at least once a month, particularly in the fall, when mite pressure increases from natural population growth, shrinking of the brood nest, and invasion from neighboring colonies. This is also the time when bees kept in northern locations switch from the production of short-lived summer bees to long-lived winter bees, and so heavy parasitism of the developing winter bees will increase the risk that colonies will die even if mites are controlled after these bees emerge.

So if for some conditions, such as *Varroa*, disease load is more informative than prevalence, it is the opposite for diseases that are highly contagious or swift acting. AFB is highly contagious and persistent, and so some state laws require total destruction of diseased colonies even if the colony has just one AFB scale present. This makes sense when one considers that one AFB scale can contain several billion spores, that these spores remain infectious for at least 50 years, and it only takes less than 10 spores to kill a larval bee if it was fed the spores in the first day of its larval life ($LD_{50} = 8.49$, Brødsgaard et al. 1998). So in the case of AFB, a disease load of 1 scale is a sufficient threshold to implement control strategies.

Illustration 2. The National Honey Bee Disease Survey (NHBDS), funded by USDA APHIS (Animal and Plant Health Inspection Service), is an example of surveillance program designed to ensure early detection of invasive pests (i.e., *Tropilaelaps clareae* mites) which are not presently found in the USA. At the same time, this survey effort provides an opportunity to describe the prevalence and load of pathogens and parasites across the country and over time. One parasite monitored was the *Varroa* mite. In Fig. 2, the blue line represents the trend of *Varroa* mite loads in the USA throughout the 2012–2013 seasons. This graph shows the cyclic nature of the infestation loads, with a peak in end of summer to fall. In this context, let us imagine a beekeeper monitoring mites who observes 4.5 mites per 100 bees. Such a load is rather high, very close to the threshold of 5 mites per 100 bees which is sometimes referred to as the damaging level of mites in a colony. However, depending on the time of the year the interpretation will vary considerably. If the sample was taken in October, the NHBDS trend curve allows us to compare this single result to an estimated average load of 10 mites per 100 bees in the USA population at that time. This does not let us know if that level is acceptable, as the survey did not collect information about survivorship of those colonies, but it allows us to say that this beekeeper's sample would be below the norm. However, if the sample was taken in May, though it is still below the same threshold of 5 mites per 100 bees, we can see that it is far above the levels reported in the USA for that time of the year. Using only a threshold criterion would have failed to detect this anomaly, while comparing it to a descriptive study of *Varroa* loads in the USA gave us a more complete and useful story.

Box 2: Sensitivity and specificity
Whenever a test is performed to identify a disease (presence or absence), there is a certain risk of error in the diagnostic. Sometimes the test will fail to identify the presence of the disease (false negative), or sometimes the test will incorrectly detect the presence of the disease (false positive). A good test method should minimize those errors. How good a test is, is quantified as the sensitivity and specificity of a diagnostic test.

The **sensitivity** of a test is its ability to correctly identify samples with the disease. A highly sensitive test minimizes false positives. In other words, if a highly sensitive test identifies a sample as negative, we are nearly certain it is indeed negative (disease free).

The **specificity** of a test is its ability to correctly identify samples without the disease. A highly specific test minimizes false negatives. In other words, if a highly specific test identifies a sample as positive, we are nearly certain it is indeed positive (diseased).

Ideally, we would want all our diagnostic tests to be both highly sensitive and specific, but that is usually something of a tradeoff.

b. Identification of risk factors in honey bee health

While descriptive studies explain the prevalence and load of disease, analytical studies aim to identify and quantify the effects of exposure variables (or **risk factors**) on the prevalence of disease. Typically, epidemiologists look for association between exposure to a risk factor and a disease outcome by comparing populations with different exposures and see how they fare in respect to the disease of interest. Once identified, modifying risk factor exposure is the corner stone of preventive programs.

Illustration 3. An example of cross-sectional study was recently completed in Argentina. Researchers quantified the *Varroa* loads in colonies and also asked beekeepers about the management practices they used. The results showed that colonies with a mite load of 3 or more mites per 100 bees were 4.9 times more likely to die over the winter (Giacobino et al. 2015) compared to colonies with mite loads below 3 mites per 100 bees. Beekeepers who did not monitor mite loads after they applied treatment, or did not requeen colonies the previous year were also more likely to experience higher rates of colony mortality.

It is important to remember that *association* (correlation) is not the same as *causation*. Most epidemiological studies are **observational** rather than experimental. This means that they take advantage of "natural experiments" in which the exposure (and sometimes the outcome) has already occurred, or occurs without the intervention of the researcher. Such "natural experiments" provide no guarantee that the two groups being compared are identical in all aspects other than the exposure/lack of exposure of interest. Experimental studies, on the other hand, try to ensure that all aspects are similar before applying the exposure themselves to a random subset of the experimental subjects. Epidemiologists strive to identify and control for all extraneous variables ("confounders") that may correlate in unexpected ways with the observed results. Whenever possible, confounders are accounted for when analyzing results from observational studies, because, if unchecked, they can bias the interpretation of the results. Even when cofounding variables are identified and controlled, scientists rarely identify risk factors as "causal" of an outcome without experiment-based evidence of these associations. Observational studies can also serve as a basis for identifying the origin (etiology) of new problems and for helping to formulate hypothesis for later experimental testing.

Illustration 4. A recent study set out to document risk factors associated with increased risk of colony mortality in three migratory beekeeping operations (vanEngelsdorp et al. 2013). The researchers found that "queen events" (evidence of a queen replacement or queen failure) was associated with an increased risk of colony death in the short term (~ 50 days following the event). In this study, "queen events" were the exposure variable of interest, and "colony death" the disease or outcome of interest. This is an example of observational (non-experimental) study as the researcher did not induce any of the queen events to follow their impact on the mortality rate. Instead, they took advantage of natural events, carefully recorded, to get insights into honey bee health mechanisms.

Measures of association
When designing studies, epidemiologists plan how they will select the subjects, follow them over time, and analyze their results. This is referred to as the "study design." There are many different study designs meant to identify possible associations between exposure and disease outcome. Three of the most widely used of these study designs include *cohort* studies, *case-control* studies, and *cross-sectional* studies (and are illustrated and explained in Figs. 3, 4 and 5).

Each study design has its strengths and weaknesses, and a detailed explanation of these differences is beyond the scope of this chapter. At the core, these differences revolve around how the subjects of the study are selected (either based on their disease or exposure status), which affects how results should be interpreted. Either way, they compare the "risk" (or probability) of disease occurring in two different groups—one exposed to the risk factor and the other not—from the same population. The results are usually presented as relative risk (also called **risk ratio**, RR) or relative odds (also called **odds ratio**, OR) (see Figs. 3, 4 and 5 for details).

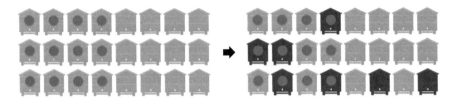

Fig. 3 Fictitious cohort study. *Legend Green* colonies = disease absent; *Red* colonies = disease present; *Dot* = exposure present (before the start of the study) to a certain factor X. The *arrow* represents the passage of time. In a cohort study, a group of disease-free subjects (the cohort) is selected based on their exposure status (both exposed and non-exposed) to a risk factor of interest (*left panel*). All of the hives would then be followed for a set period of time, and the incidence of disease in both the exposed and unexposed subgroups are monitored (*right panel*). This kind of study allows for the calculation of relative risk (RR) which is the ratio of incidence of disease (or risk) in the exposed population (R_{Exp}) divided by the incidence of disease in the unexposed population (R_{NExp}). In other words, it is a measure of the increased (or decreased) risk subjects have of developing a disease after being exposed to a risk factor. For this example, the probability, or risk, of a colony in the exposed group (*with dots in the figure*) to develop the disease during the study period would be 0.42 (5 on 12). Not all colonies exposed will develop the disease. This rate should be compared to the risk for colonies without a known exposure (*without dots in the figure*) to develop the problem, which is about 0.17 (2 on 12). In this example, the disease does also occur in colonies that were not exposed to the risk, but at much lower rate than for exposed colonies. The relative risk is calculated at 2.47 (RR = R_{Exp}/ R_{NExp} = 0.42/0.17), which represents an increased risk of 147% ((2.47−1) × 100). This means that exposed colonies were 147% more likely to become diseased than non-exposed colonies. The conclusion is that colonies exposed to the product X present a higher risk of developing the disease than the non-exposed colonies, and the recommendation would be that beekeepers avoid the use of product X

A RR of OR of 1 indicates that both groups show similar risks of disease, irrespective of the level of the exposure. So the exposure seems unassociated with the disease. A RR or OR greater than 1 indicates that the group exposed shows higher levels of disease, so the exposure is associated with the disease. A RR or OR less than 1 indicates that the group exposed shows lower levels of disease, which suggest the exposure reduced disease prevalence. The greater the magnitude of the difference, the greater the "strength of the association": The more one group shows an increased risk for the disease compared to the other group.

Illustration 4 (continued). The study of migratory operations is an example of cohort study where groups with different exposure histories (queen events) were followed and compared in terms of disease incidence (in this case, colony mortality). A total of 284 inspections were performed, from which 35 showed signs of a queen event. The colonies were inspected again after ~50 days to determine the outcome status for the whole colony. The table of incidences is shown below. From it, we can determine that colonies that underwent a queen event showed a risk of 0.31 or 31% (R_{Exp} = 11/35) of

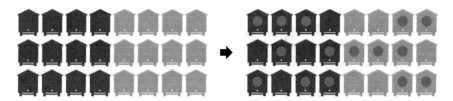

Fig. 4 Fictitious case-control study. *Legend Green* colonies = disease absent; *Red* colonies = disease present; *Dot* = exposure present (before the start of the study) to a certain factor X under study. The *arrow* represents the passage of time. In a case-control study, a group of diseased subjects (the cases) are compared to a group of disease-free subjects (the controls) (*left panel*). Ideally, control subjects resemble the disease subjects as closely as possible. Their history is then compared (usually through surveys, or tests are performed) to establish which of them were exposed to the risk factor under study (*right panel*). Because the proportion of cases to control is unlikely to be representative of their proportion in the source populations (we actively looked for the diseased colonies), it would not be fair to calculate totals, probabilities, and risk ratios. However, we can compare odds: The probability that some event will occur compared to the probability that it will not occur. Odds ratios (OR) are tricky and easily misinterpreted, even by professionals. For this example, of the 24 colonies selected based on their disease status (12 diseased and 12 controls non-diseased), 15 of the colonies were found to have been exposed to the factor X under study (*with dots in the figure*). The odds of an exposed colony being a case are 8 to 7. The odds of a non-exposed colony being a case are 4 to 5. The odds ratio (OR) would therefore be 1.4 (8/7 divided by 4/5). We would be interpreted as a 40% increase of odds of developing the disease in the exposed population compared to the non-exposed population. When the disease is uncommon, OR will be reasonably good estimates of risk ratio and can be interpreted similarly

dying over the next $\sim$ 50 days. Colonies that did not experience a queen event showed a risk of 0.10 or 10% (R_{NExp} = 25/249) of dying over the next 50 days. In this study, the relative risk is calculated at 3.1 ($RR = R_{Exp}/R_{NExp}$ = 0.31/0.10), which represents an increased risk of 210% (($3.1-1$) × 100). This means that the risk of dying for colonies who experienced a queen event was more than two times more likely to die than those colonies that did not experience a queen event (vanEngelsdorp et al. 2013).

Table of incidences from Illustration 4 cohort

		Outcome		
		Dead	Alive	Total
Queen event	Yes	11	24	35
	No	25	224	249
	Total	36	248	284

It is critical to remember that association is not causation, so it would be incorrect to say that the queen event caused the increase in colony loss. Based on this study, it is not possible to determine if that queen events caused the increased mortality, or if some other factor caused colonies to die also caused an increase in queen events.

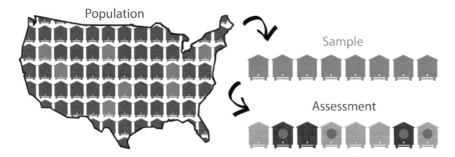

Fig. 5 Basic design of a cross-sectional study. In cross-sectional studies, subjects are first selected according to a particular sampling scheme (random, convenience or other), without regard for their disease or exposure status. They are referred to as the **sample**. Then both exposure status and outcome status are assessed at the same time. Sometimes, the investigators try to determine past levels of exposure through retrospective surveys. For instance, they might be able to glean significant information from beekeeping records. Those studies have the advantage that both outcome and exposure levels are representative of their true prevalence in the target population (subjects exposed and/or diseased are not more likely to be selected). Another variant is to follow the same population over time in a series of snapshots of cross-sectional studies. While **cohorts** start with disease-free subjects and investigate *incidence* of problems (i.e., development of new cases), **cross-sectional** studies focus on the *prevalence* within (i.e., the existing cases) a population at the time of the study. Therefore, in cross-sectional studies, the measure of risk is based on the prevalence of disease outcome in groups that have had or have not had an exposure to the risk factor of interest. This is expressed as a relative risk (RR) and is calculated exactly like the relative risk in Cohort studies. The difference, however, is that in this case relative risk relates to the risk of having the disease rather than the risk of developing it. This is an important distinction when interpreting the results: If a factor improves the survivorship of diseased colonies compared to non-exposed diseased colonies (but without curing them), it could be misinterpreted as being associated with the disease, because most disease colonies still alive would be most probably exposed (the others being already lost)

2 Significance of Epidemiology for Your Beekeeping Management

Before collecting any data, epidemiologists plan their experiments and decide which exposures and outcomes they will investigate. This is because the real world is complex. Multiple causes can exist for almost every outcome and every exposure variable can affect many different diseases (Dohoo et al. 2003). Epidemiologists have to focus on a specific problem. Many times even minor unrecognized factors can dramatically impact outcomes. Colonies managed by different beekeepers will be subjected to very different regimens (equipment, feeding, treatment, migration…). Even within the same beekeeping operations, apiaries will differ between each other in terms of availability of resources. Further within the same apiary, colonies can experience very different microclimates (for instance, some colonies are predominantly in the shade while others in the sun). A careful study would try to control these extraneous variables; for instance making sure all apiaries were all in full sun, so that any potentially cofounding effects are minimized.

Epidemiologists work at the population level, trying to estimate the difference the implementation of preventive or curative practices would have for the whole population. In some respects, epidemiologists ask "what if." What if the risk factor associated with a disease was removed? How many fewer cases of the disease could we expect?

As epidemiologists deal with calculating the chance of something occurring or not, they cannot make predictions for individuals, rather they can make predictions at the population level. Thus, a large part of epidemiology involves the application of risk easement strategies in order to reduce disease prevalence for the whole population.

There is a common saying: "a poll is only as good as its sample size." The same holds true for epidemiological studies. Whether the interest is in knowing the prevalence of a disease in a population or the strength of its association to an exposure, it is important that the sample is **representative** of the overall population. Populations have variability, and a good sample has the same variability. Epidemiologists usually convey this idea with a measure of uncertainty around their results, such as the **confidence intervals** (CI). Usually, the greater the sample size (the more subjects in the study), the smaller the CI around the estimate (the smaller the incertitude).

Illustration 2 (continued). The US National Honey Bee Disease Survey report (Traynor et al. 2016), summarizing the results from 2009 to 2014, indicated that migratory beekeepers had significantly lower *Varroa* prevalence than stationary operations (84.9% [81.4–87.8%] versus 97.0% [95.6–97.9%]). The estimates are followed by a bracket indicating the breadth of the confidence interval. Because those two intervals (the ones for stationary and for migratory) do not overlap, we are confident in saying their prevalence are significantly different.

Traditionally, statisticians employ a "95% CI," which indicates that, if we were to repeat the study 100 times, with 100 samples drawn randomly from the same population, and that a CI was calculated for each trial, 95 of those CI would contain the population's true *Varroa* prevalence.

3 Current State of Honey Bee Colony Population and Health

There are many ways to monitor honey bee health. One measure is the total numbers of managed honey bee colonies over time. Honey bee populations have increased globally by 64.7% since 1961, reaching a total of 81 million managed honey bee colonies in 2013 (Food and Agriculture Organization of the United Nations (FAO) 2015). This global increase is largely driven by increases in colony numbers in some regions of the world (Asia and South America) which masks significant decreases experienced in other regions, such as that documented in

Europe (−20.3%, Potts et al. 2010a) and the USA (−52.1%, vanEngelsdorp and Meixner 2010) (see Fig. 6). While total colony counts are good indicators of managed pollinator availability, they inadequately represent honey bee health. Managed honey bee colony population trends are mostly driven by socioeconomic factors (such as number of beekeepers, price of honey, political disruption) (Aizen and Harder 2009; Potts et al. 2010a, b; vanEngelsdorp and Meixner 2010) rather than biological. Total colony counts, estimated once a year, ignore the beekeeper practice of replacing dead-outs to keep operational numbers up. Beekeepers divide healthy colonies and/or buy and install packages in order to replace dead-out colonies or to increase operational size, so that the absolute number of colonies can be stable or even increasing year after year, even if colonies are subjected to high mortality rates (vanEngelsdorp et al. 2007).

Because of the ability to replace dead-out colonies quickly, which is particular to managed systems (as opposed to wild pollinators), honey bee health is better represented by measuring the rate of colony mortality over a defined time frame. In 2008, the COLOSS (prevention of honey bee Colony LOSSes) network—formed of honey bee experts from Europe, North America, and some other regions around the world—developed a standardized questionnaire to gather information about colony losses in an effort to enable comparison between participating countries (van der Zee et al. 2012). While at first these survey efforts focused on winter mortalities,

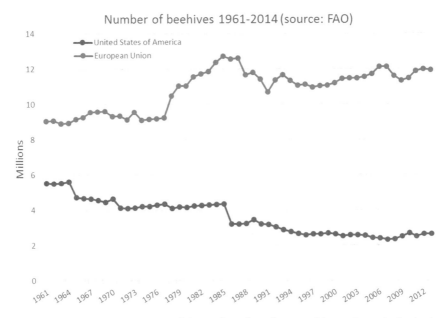

Fig. 6 Population trend. Estimates of the total number of managed honey bee colonies in the USA and European Union between 1961 and 2014 (Food and Agriculture Organization of the United Nations (FAO) 2015). © FAO 2015 Production/Live Animals/Beehives. This is an adaptation of an original work by FAO. Views and opinions expressed in the adaptation are the sole responsibility of the author of the adaptation and are not endorsed by FAO

more recent USA efforts have included calculating summer loss rates as well. It has long been assumed that summer loss rates are minimal; however, survey efforts have shown that in the USA summer losses are not negligible (Steinhauer et al. 2014) and so should also be considered when attempting to describe the status of honey bee health.

Over the last 10 years, the rate of honey bee losses over the winter in the USA has ranged from 22.3% to 35.8%, averaging around 28%. Over the 6 years for which summer (as defined by the period between April and October) numbers are available, summer losses ranged from 16.2% to 25.3% and averaged 21% (vanEngelsdorp et al. 2007, 2008, 2010, 2011, 2012; Spleen et al. 2013; Steinhauer et al. 2014; Lee et al. 2015b; Seitz et al. 2015; Kulhanek et al. 2017). Those loss estimates are far above the levels beekeepers themselves judge acceptable (16%, average of 10 years).

The causes of high levels of managed honey bee colony losses are multiple and probably interacting (Potts et al. 2010b). Honey bees face a very diverse array of threats (reviewed in Potts et al. 2010b; vanEngelsdorp and Meixner 2010) from diseases and parasites to reduced quality and quantity of bee forage due to land-use change, climate change, contaminations by pesticides (applied both outside and inside the hive) and, at least for USA populations, potential loss of genetic variability (but see Wallberg et al. 2014).

High levels of colony loss throughout the year seriously threaten the sustainability of beekeeping operations. Replacing dead colonies is costly, both directly (e.g., purchase of queens and bees) and indirectly, resulting from reduced productivity of split colonies. Weak and unhealthy colonies are also more costly to maintain as they need more feed, more frequent inspection and disease treatments. Weaker colonies also do not generate the same return as healthy strong colonies. Almond producers commonly have provisions in their pollination contracts that pay premiums for strong colonies while enforcing penalties for weak colonies. In fact some pricing schedules are now based on frame counts instead of the number of hives (Champetier 2011).

4 Summary

Epidemiology emphasizes that health is a common good. By focusing research on health at the population level, epidemiology can make a great impact in improving health. Large-scale epidemiological studies are important both to produce reliable accounts of the status of honey bee health and also to react efficiently to abnormal health events, develop and test hypotheses on disease etiology and to inform prevention and control strategies.

The same key principles that make epidemiological studies successful at population levels apply at apiary or operational levels and should be applied by every beekeeper. These recommendations include:

1. Carefully apply the preventive recommendations developed locally.
2. Monitor disease levels as often as practical throughout the year.
3. Compare the levels present in your own apiary to a quality baseline to detect abnormalities.
4. Apply the proper recommended control strategies when problems are detected.
5. Assess the efficacy of such control methods whenever they are used. This means always doing a recheck for the problem to be sure the control method(s) used were effective.
6. Always keep quality records and when possible participate in national surveys.

References

Aizen MA, Harder LD (2009) The global stock of domesticated honey bees is growing slower than agricultural demand for pollination. Curr Biol 19(11):915–918. doi:10.1016/j.cub.2009.03.071

Brødsgaard CJ, Ritter W, Hansen H (1998) Response of in vitro reared honey bee larvae to various doses of Paenibacillus larvae larvae spores. Apidologie 29(6):569–578. doi:10.1051/apido:19980609

Burgess AF, Howard LO (1906) The Laws in force against injurious insects and foul brood in the United States. Washington, D.C. : U.S. Department of Agriculture, Bureau of Entomology, 236 pp

Champetier A (2011) The foraging economics of honey bees in almonds. Presented at the Agricultural & Applied Economics Association's 2011 AAEA & NAREA Joint Annual Meeting, University of California, Davis, Pittsburgh, Pennsylvania

Dohoo IR, Martin SW, Stryhn H (2003) Veterinary epidemiologic research. 706 pp, AVC Inc., Charlottetown

Food and Agriculture Organization of the United Nations (FAO) (2015) FAOSTAT [WWW Document]. http://faostat3.fao.org/. Accessed 13 Aug 2015

Frey E, Rosenkranz P (2014) Autumn invasion rates of Varroa destructor (Mesostigmata: Varroidae) into honey bee (Hymenoptera: Apidae) colonies and the resulting increase in mite populations. J Econ Entomol 107(2):508–515. doi:10.1603/EC13381

Giacobino A, Molineri A, Cagnolo NB et al (2015) Risk factors associated with failures of Varroa treatments in honey bee colonies without broodless period. Apidologie 1–10. Doi:10.1007/s13592-015-0347-0

Kulhanek K, Steinhauer N, Rennich K, Caron DM, Sagili RR, Pettis JS, Ellis JD, Wilson ME, Wilkes JT, Tarpy DR, Rose R, Lee K, Rangel J, vanEngelsdorp D (2017) A national survey of managed honey bee 2015–2016 annual colony losses in the USA. J Apic Res 56(4):328–340. DOI: 10.1080/00218839.2017.1344496

Laurent M, Hendrikx P, Ribiere-Chabert M et al (2015) A pan-European epidemiological study on honeybee colony losses 2012–2014. European Union Reference Laboratory for honeybee health (EURL)

Lee K, Steinhauer N, Travis DA et al (2015a) Honey bee surveillance: a tool for understanding and improving honey bee health. Curr Opin Insect Sci 10:37–44. Doi:10.1016/j.cois.2015.04.009

Lee KV, Steinhauer N, Rennich K et al (2015b) A national survey of managed honey bee 2013–2014 annual colony losses in the USA. Apidologie 1–14. Doi:10.1007/s13592-015-0356-z

Lindström A, Korpela S, Fries I (2008a) The distribution of Paenibacillus larvae spores in adult bees and honey and larval mortality, following the addition of American foulbrood diseased brood or spore-contaminated honey in honey bee (Apis mellifera) colonies. J Invertebr Pathol 99(1):82–86. doi:10.1016/j.jip.2008.06.010

Lindström A, Korpela S, Fries I (2008b) Horizontal transmission of Paenibacillus larvae spores between honey bee (Apis mellifera) colonies through robbing. Apidologie 39(5):515–522. doi:10.1051/apido:2008032

Pettis JS, Rose R, Lichtenberg EM, Chantawannakul P, Buawangpong N, Somana W, Sukumalanand P, Vanengelsdorp D (2013) A rapid survey technique for Tropilaelaps mite (Mesostigmata: Laelapidae) detection. J Econ Entomol 106(4):1535–1544. doi:10.1603/EC12339

Potts S, Roberts S, Dean R, Marris G, Brown M, Jones R, Neumann P, Settele J (2010a) Declines of managed honey bees and beekeepers in Europe. J Apic Res 49(1):15. doi:10.3896/IBRA.1.49.1.02

Potts SG, Biesmeijer JC, Kremen C, Neumann P, Schweiger O, Kunin WE (2010b) Global pollinator declines: trends, impacts and drivers. Trends Ecol Evol 25(6):345–353. doi:10.1016/j.tree.2010.01.007

Seitz N, Traynor KS, Steinhauer N, Rennich K, Wilson ME, Ellis JD, Rose R, Tarpy DR, Sagili RR, Caron DM, Delaplane KS, Rangel J, Lee K, Baylis K, Wilkes JT, Skinner JA, Pettis JS, vanEngelsdorp D (2015) A national survey of managed honey bee 2014–2015 annual colony losses in the USA. J Apic Res 54(4):292–304. doi:10.1080/00218839.2016.1153294

Spleen AM, Lengerich EJ, Rennich K, Caron D, Rose R, Pettis JS, Henson M, Wilkes JT, Wilson M, Stitzinger J, Lee K, Andree M, Snyder R, vanEngelsdorp D (2013) A national survey of managed honey bee 2011–12 winter colony losses in the United States: results from the bee informed partnership. J Apic Res 52(2):44–53. doi:10.3896/IBRA.1.52.2.07

Steinhauer NA, Rennich K, Wilson ME, Caron DM, Lengerich EJ, Pettis JS, Rose R, Skinner JA, Tarpy DR, Wilkes JT, vanEngelsdorp D (2014) A national survey of managed honey bee 2012–2013 annual colony losses in the USA: results from the bee informed partnership. J Apic Res 53(1):1–18. doi:10.3896/IBRA.1.53.1.01

Traynor KS, Rennich K, Forsgren E, Rose R, Pettis J, Kunkel G, Madella S, Evans J, Lopez D, vanEngelsdorp D (2016) Multiyear survey targeting disease incidence in US honey bees. Apidologie 47(3):325–347. doi:10.1007/s13592-016-0431-0

van der Zee R, Pisa L, Andonov S, Brodschneider R, Charriere J-D, Chlebo R, Coffey MF, Crailsheim K, Dahle B, Gajda A, Gray A, Drazic MM, Higes M, Kauko L, Kence A, Kence M, Kezic N, Kiprijanovska H, Kralj J, Kristiansen P, Martin Hernandez R, Mutinelli F, Nguyen BK, Otten C, Ozkirim A, Pernal SF, Peterson M, Ramsay G, Santrac V, Soroker V, Topolska G, Uzunov A, Vejsnaes F, Wei S, Wilkins S (2012) Managed honey bee colony losses in Canada, China, Europe, Israel and Turkey, for the winters of 2008–9 and 2009–10. J Apic Res 51(1):91–114. doi:10.3896/IBRA.1.51.1.12

vanEngelsdorp D, Meixner MD (2010) A historical review of managed honey bee populations in Europe and the United States and the factors that may affect them. J Invertebr Pathol 103:S80–S95. doi:10.1016/j.jip.2009.06.011

vanEngelsdorp D, Underwood R, Caron D, Hayes J (2007) An estimate of managed colony losses in the winter of 2006-2007: A report commissioned by the apiary inspectors of America. Am Bee J 147(7):599–603

vanEngelsdorp D, Hayes J, Underwood RM et al (2008) A survey of honey bee colony losses in the U.S., fall 2007 to spring 2008. PLoS ONE 3(12). Doi:10.1371/journal.pone.0004071

vanEngelsdorp D, Hayes J, Underwood RM, Pettis JS (2010) A survey of honey bee colony losses in the United States, fall 2008 to spring 2009. J Apic Res 49(1):7–14. doi:10.3896/IBRA.1.49.1.03

vanEngelsdorp D, Hayes J, Underwood RM, Caron D, Pettis J (2011) A survey of managed honey bee colony losses in the USA, fall 2009 to winter 2010. J Apic Res 50(1):1–10. doi:10.3896/IBRA.1.50.1.01

vanEngelsdorp D, Caron D, Hayes J, Underwood R, Henson M, Rennich K, Spleen A, Andree M, Snyder R, Lee K, Roccasecca K, Wilson M, Wilkes J, Lengerich E, Pettis J (2012) A national survey of managed honey bee 2010-11 winter colony losses in the USA: results from the Bee Informed Partnership. J Apic Res 51(1):115–124. doi:10.3896/IBRA.1.51.1.14

vanEngelsdorp D, Tarpy DR, Lengerich EJ, Pettis JS (2013) Idiopathic brood disease syndrome
    and queen events as precursors of colony mortality in migratory beekeeping operations in the
    eastern United States. Prev Vet Med 108(2–3):225–233. doi:10.1016/j.prevetmed.2012.08.004
Wallberg A, Han F, Wellhagen G, Dahle B, Kawata M, Haddad N, Simões ZLP, Allsopp MH,
    Kandemir I, De la Rúa P, Pirk CW, Webster MT (2014) A worldwide survey of genome
    sequence variation provides insight into the evolutionary history of the honeybee Apis
    mellifera. Nat Genet 46(10):1081–1088. doi:10.1038/ng.3077

Author Biographies

Nathalie Steinhauer is a Ph.D. candidate in Dennis vanEngelsdorp's lab at *University of Maryland*, Department of Entomology. She graduated with a M.S. in Biological Sciences from *Universite Libre de Bruxelles*, Belgium and with an M.Res. in Ecology, Evolution, and Conservation from *Imperial College London*, UK.

Her Ph.D. project's objectives are to apply epidemiological approaches to honey bee health and identify best management practices associated with reduced colony mortality. As part of the Bee Informed Partnership (BIP), Nathalie has been primarily responsible for organizing and leading the effort of the analyses of the National Honey Bee Colony Loss and Management Surveys for the past 4 years. She has collaborated in the publishing of the Loss Survey results in peer-reviewed journals and presented her research in numerous scientific meetings and extension events.

Find more on our wonderful team at BIP's website: www.beeinformed.org

Dennis vanEngelsdorp Dennis is an Assistant Professor at the University of Maryland has a broad interest in pollinator health. The focus of his current work involves the application of epidemiological approaches to understanding and (importantly) improving honey bee health. Dennis is the founding president of the Bee Informed Partnership (BeeInformed.org) which attempts to provide a platform to collect "big data" on the state of managed honey bee colony health. Analysis of these data is providing important insights into the role beekeeper management practices and environmental factors (such as landscape pesticides and climate) have on keeping colonies alive.

Small Hive Beetles (*Aethina Tumida* Murray) (Coleoptera: Nitidulidae)

Christian W.W. Pirk

Abstract

The natural distribution of *Aethina tumida* (SHB) is limited to sub-Saharan Africa, where SHB is considered to be a minor and negligible pest. The introduction into countries outside of Africa, like Australia, the USA, and the recent introduction into Italy, had devastating effects. This suggests that the SHB can be a serious threat to beekeeping operations when using European stock of *A. mellifera*. Nevertheless, comparative research in Africa and North America over the last decade suggests that there are only quantitative and no qualitative differences between the subspecies, which makes the one subspecies susceptible and the other one resistant. The scientific evidences indicate that general aspects like colony activity levels may play a crucial role in resistance. Therefore, changes in the management practice can significantly decrease the susceptibility of colonies toward SHB. The rule of thumb is to help the colonies to help themselves by maintaining strong colonies, ensuring workers have access to all parts of the hive where beetles hide and/or reproduce. Preventative measures should also be put in place so SHBs are unable to reproduce outside the hive, in the honeyroom or in old, stored comb.

The recent detection in 2014 of small hive beetles (SHBs) (*Aethina tumida* Murray) in Italy triggered serious concern within the agricultural and apicultural communities in Europe (Mutinelli et al. 2014). The beetle is native to sub-Saharan Africa, and its distribution has extended to *Apis mellifera* colonies of European origin, in

C.W.W. Pirk (✉)
Social Insects Research Group, Department of Zoology and Entomology, University of Pretoria, Private Bag X20, Hatfield, 0028 Pretoria, South Africa
e-mail: cwwpirk@zoology.up.ac.za

© Springer International Publishing AG 2017
R.H. Vreeland and D. Sammataro (eds.), *Beekeeping – From Science to Practice*,
DOI 10.1007/978-3-319-60637-8_9

143

far flung areas such as the USA and Australia (Hepburn and Radloff 1998; Neumann et al. 2016). Although SHB can be a serious threat to beekeeping operations using European-derived stock of *Apis mellifera,* it is considered only a minor pest (Figs. 1–6) within its native range (Hepburn and Radloff 1998; Pirk et al. 2014, 2015). The range of SHB in sub-Saharan Africa is transitory to a certain extent due to beekeeping activities; South Africa has an apicultural industry similar to that of Europe (Dietemann et al. 2009). Indeed, SHB may be perceived as one of the contributing causes of colony losses in South Africa (Pirk et al. 2014), and the economic impact is perceived as being rather low (Johannsmeier 2001). Nevertheless, the expanding global distribution of SHB, starting in 1996 in Charleston, South Carolina, USA to the Philippines in 2014 (reviewed in Neumann et al. 2016), makes it a more and more important apicultural topic.

The Lifecycle of Small Hive Beetles—Opportunities to Control the Pest

Significant sections of the life cycle of the SHB take place within the colony, with the remaining portions taking place outside the hive. Adult small hive beetles locate a suitable colony and try to enter; when successful, it is chased by the workers and it will hide in cracks and gaps which the bees are not able to access (Fig. 7). In these cracks, other beetles will accumulate as well and mating can be frequently observed. Mated females then lay eggs in cracks, gaps and even under the cell cappings of sealed honeybee brood. The eggs hatch after 1–3 days, and the larvae start feeding on honey and pollen stores as well as honeybee brood. The duration of

Fig. 1 Honeybee workers attacking a small hive beetle (*Photo* C. Laing)

Fig. 2 Small hive beetle being nearly entombed in a crack in an *Apis mellifera scutellata* colony (*Photo* C. Laing)

Fig. 3 Two small hive beetles coercing a honeybee worker for food (*Photo* C.W.W. Pirk)

Fig. 4 Small hive beetles trying to hide in the cracks of a hive lid (*Photo* C. Laing)

Fig. 5 Small hive beetles, guarded by bees, hiding at the bottom of empty cells (*Photo* C.W.W. Pirk)

the feeding stage depends on the diet the larvae consume; after 3–21 days, the larvae transition into the wandering phase during which the larvae aggregate, become positively phototactic, and leave the hive to pupate in suitable soil. The pupal stage can last anywhere from a few weeks to a month depending on the

Fig. 6 Entrance of an *A. m. scutellata* colony, narrowed using propolis (*Photo* C. Laing)

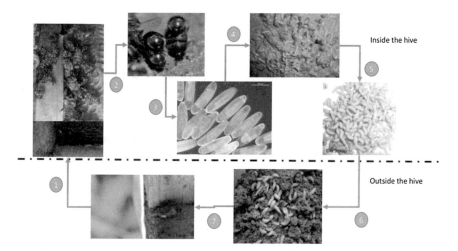

Fig. 7 Life cycle of the small hive beetle. Adult beetles locate potential hive and sneak into it (*1*). They hide in cracks and gaps, and sometimes they are guarded by bees. In the amalgamations of beetles mating takes place (*2*), perhaps even before entering the hive. Female SHB lays eggs (*3*), and when the larvae hatch, they can cause significant damage (*4*). After the feeding stage, the larvae change into the "wandering phase" (*5*) and leave the hive to pupate in the soil (*6*). After a few weeks, the next generation of SHB emerges from the soil

temperature it is exposed to. Normally, after 14-21 days the new generation of small hive beetles emerges from the soil and after a short maturation period the beetles are ready to invade the next hive.

During the SHB life, several opportunities arise during which control of this honeybee pest can be implemented. Unlike the Varroa mite (*Varroa destructor* Anderson and Trueman 2000) which spends its life almost exclusively within the hive, SHB periodically leaves the hive (Lundie 1940; Schmolke 1974; Neumann and Elzen 2004; Neumann and Ellis 2008). Firstly, an adult beetle has to locate a hive, and the next step is to enter the hive without being detected and attacked by the guard bees. While within the colony, honeybee workers are able to detect SHB and attack it; therefore, the beetle has to find shelter in the gaps and cracks (Fig. 4) where it can safely hide from the fierce mandibles of the workers. Honeybee workers are able to decapitate SHB but usually the beetles are well protected by their tough exoskeleton (Elzen et al. 2001; Neumann et al. 2001). Once the beetle has found a crack in which to hide, the next problem faced is that the honeybee workers will stand guard to keep the beetles away from the stores and brood on the comb (Fig. 5). To further exacerbate matters for the beetles, the workers may begin to construct prisons using wax and propolis (Fig. 2), which help to contain the beetles (Neumann et al. 2001; Ellis et al. 2002b, 2003b).

During their imprisonment, the beetles and bees interact; SHBs have an innate ability to trick the guard bees into feeding them (Fig. 3), while in prison (Ellis et al. 2002b), the begging behavior of the beetles seems to be evolving over time (Neumann et al. 2015).

From the beetle perspective, the advantage of the being in a confined area is that it is most likely a sanctuary for other SHB as well; therefore, finding a mate is made easy. Once the female has mated, she has to find a place to lay her eggs. Laying her eggs in the prisons or cracks in the hive is not ideal as these may be located too far from the food sources of honey, pollen, and honeybee brood. The SHB larvae are also easily recognized by the workers, swiftly removed and dropped outside the colony (Schmolke 1974; Neumann and Härtel 2004). In a weak colony, the number of guard bees may be reduced, enabling SHB females to enter areas where brood, pollen, and honey combs are stored, or parts of the combs which are not accessible to the bees for cleaning. Such areas provide oviposition sites for SHB females. Moreover, Ellis et al (2003c) reported that females are able to lay eggs on the inside lid of sealed cells. The SHB larvae tunnel through cells containing honey, pollen, and brood, and as a result the structure of the comb is compromised and slowly disintegrates and collapses; the excess honey drips out of the cells and begins to ferment. Dripping fermenting honey combined with SHB larvae excreta results in conditions which repel bees and creates a larger area that the workers will not patrol.

After around two weeks of feeding, the SHB larvae develop into the so-called wandering phase and cease feeding. These wandering larvae now have to pupate in the soil and exhibit phototactic behavior, exactly the opposite behavior of the newly emerged larvae. This distinct behavior is an ideal method to discriminate between the feeding and the wandering stage of the larvae. When opening a hive box, larvae in the feeding stage, when exposed, would drop from the frames deeper into the darker areas of the hive. The negative phototaxy of the feeding stage could be utilized as a within hive control mechanism if one would be able to create a light cue within the dark hive environment, which chases the larvae in darker areas containing a trapping mechanism. However, exposing the bees to a light cue might have other disadvantages. The positive phototaxy of the wandering stage could be utilized for light trapping; however, this would only work for population control, as the damage of the feeding stage would have already been done to the colony. Therefore, the strong phototaxy could be explored for control and monitoring tools.

Normally, the larvae leave the hive through the entrance and as soon as it comes into contact with soil, it starts burrowing (reviewed in Neumann et al. 2016, Fig. 7).

After a month or so of pupation in the soil, a new generation of SHBs emerges, and after a maturation period (personal observations), they are ready to find a hive to invade.

Small hive beetles are less of a problem in the northern parts of South Africa than wax moth (*Galleria mellonella* L.) (Strauss et al. 2013), and yet dealing with SHBs are as difficult as dealing with wax moth; there are, however, a few rules one has to follow.

With only some minor adjustments to beekeeping practice, it is possible to keep bees in the areas where SHBs naturally occur. Based on general good beekeeping principles which include good sanitation and hygiene in the honey house, apiary and storage rooms, thereby preventing/suppressing SHB reproduction and removing resources which SHBs utilize to complete their life cycle. During the honey harvest process, the combs should immediately be either stored at 4 °C and below or in sealed SHB-proof containers. Also the storage of any old combs, even with a small amount of pollen or honey stores present, should be avoided. However, if it is necessary to save the old combs, these should be stored in sealed containers at low temperatures, but it is advisable/preferable not to store any old combs at all.

Apart from these basic beekeeping techniques, various traps are available to reduce the number of SHBs within the colony. For example, the "Schäfer trap" (Schäfer et al. 2008) comprises a piece of corrugated plastic (75 × 500 × 4 mm) creating rows of narrow tunnels, which is placed on the bottom board at the entrance of the hive. The beetle takes refuge in these tunnels and, as it is not necessary to open the hive, the trap is easily removed during inspection, cleaned, and redeployed as frequently as required.

Other traps are the Beetle Barn™ (Rossmann Apiaries), the Hood Trap™, and the AJ's Beetle Eater™ trap. Something all these traps have in common is that they provide SHB hiding places in areas which are inaccessible to the bees. These traps can be used in combination with a chemical application ensuring that the beetles remain in the traps once they have entered. It is important to bear in mind possible knock-on effects any chemicals used may have on bees and bee products.

In a study by Bernier et al. (2015), significant variation in the trapping success of the three tested traps was shown. It is, therefore, important to be aware of the advantages and disadvantages of the different trap types. All three traps tested by Bernier et al. (2015) catch SHBs and so does the Schäfer trap (Schäfer et al. 2008). However, only Schäfer et al (2008) evaluated the efficiency of the trap, which was between 40 and 50% of the beetles present in the hive being trapped. Bernier et al. (2015) compared the number of trapped beetles using the AJ's Beetle Eater™ (AJ's Beetle Eater), Beetle Barn™ (Rossmann Apiaries), and Hood Trap™ (Brushy Mountain Bee Farm). The Beetle Eater had the highest numbers of trapped beetles followed by the Hood Trap™ and the Beetle Barn™. Also the positioning of the trap inside the hive might affect the efficiency, for reasons unknown the AJ's Beetle Eater™ trap on the left side caught more than on the right side of the hive box. Furthermore, for the Beetle Eater and the Hood Trap™, one has to open the colony and fill the traps with mineral oils and vinegar, whereas the beetle barn uses commercial pesticides, which raises the problem of potential resistance development in SHBs. All three of the traps are reduced in efficiency when the workers are allowed to close the entrances of the traps with propolis (Bernier et al. 2015). Besides the four mentioned designs above, there are various other traps of different designs, of varying efficiency and varying labor intensity, but all the traps have in common that they only control SHBs and do not fully remove them from the hives.

Comparative research as to the possible reason(s), why sub-Saharan honeybee populations are less affected by SHBs and other honeybee populations are so susceptible have been conducted over the past 20 years. Behavioral experiments conducted in Grahamstown, South Africa and in Umatilla, Florida, USA, revealed quantitative differences in the aggression levels of worker honeybees toward SHBs but not qualitative ones (Elzen et al. 2001). These quantitative differences indicate that the beetle is actually recognized as a threat by European honeybees and not simply ignored. Other behavioral mechanisms employed by African honeybee colonies when dealing with SHBs, like prison building (Neumann et al. 2001; Ellis et al. 2003c), and being coerced into feeding the beetles (Ellis et al. 2002b) (Fig. 3), were also observed in European colonies (Ellis et al. 2003c). Small hive beetle presence can even trigger absconding behavior in colonies of European honeybees (Ellis et al. 2003d), so-called nonreproductive swarming, as a reaction to unfavorable conditions in the nest (Hepburn 1988; Allsopp and Hepburn 1997; Villa 2004; Spiewok and Neumann 2006; Spiewok et al. 2006). When looking into absconding behavior of the African and European honeybees, one can find various similarities and differences. When African and European colonies were artificially infested with SHBs, their absconding behavior was very similar (Ellis et al. 2003d). The artificial infestation was done by adding 100 adult beetles per day to each of the experimental colonies and recording the amount of sealed brood, adult workers, stored pollen, and flight activity (Ellis et al. 2003e). Although there were no differences in the absconding events themselves, the two populations showed slightly different behavior in the preparation for the upcoming absconding event (Fig. 8). The African honeybees significantly reduced their pollen stores, but maintained the same sealed brood area, number of adults, and the flight activity as the control colonies (no SHB added). However, the European bees, in their preparation to abscond, decreased the sealed brood area, the number of adults also decreased, and they were less active, and yet the pollen stores remained the same (Ellis et al. 2003e). As a result, the European colonies had less workers to protect the pollen stores against SHBs, which may be the reason as to why SHBs are able to reproduce more easily in European honeybee colonies. In comparison, African honeybee workers carried on with their duties, thereby ensuring that a sufficient number of workers are present and able to deal with the beetles. Easier access by SHBs to the pollen stores due to a smaller number of workers patrolling and protecting the stores may result in the following positive feedback loop. It has been suggested that SHB facilitates the spread of the yeast (*Kodamaea ohmeri*) onto pollen, and the inoculated pollen then releases volatiles, which attract more SHBs (Torto et al. 2007). Hence, a colony that does not defend the pollen stores and allows SHBs to access the stores will have more yeast on the pollen stores. The more yeast present on the pollen, the more volatiles are released, attracting even more beetles and

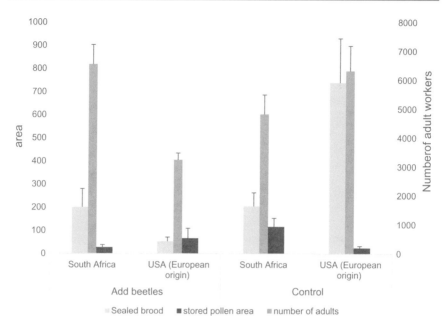

Fig. 8 Results from the Ellis et al. (2003e) study; the one group of hives was treated by adding SHBs on a daily basis and the other group served as a control. After 15 days, the total sealed brood area (cm^2), total pollen area (cm^2), and number of adults were counted. The experiment was once done in South Africa using *Apis mellifera capensis* and a second time in the USA using *Apis mellifera* of mixed European origin

subsequently increasing the pressure on the colony. Again, this is not specific to European colonies but also occurs within the SHB's native range, and this alone cannot explain the differences in susceptibility between the African and European honeybee populations. Also the potential of these volatiles to lure SHBs into traps could be explored.

In a recent study by Pirk and Neumann (2013), the level of activity of honeybee workers was investigated and how it affected their interactions with SHBs. Interactions of young (<24 h old) and older honeybee workers (<7 days old) and mature adult SHB were tested and recorded. The results clearly show that activity levels of both the young and older workers did not differ; therefore the age of the honeybee does not seem to play a significant role in imprisoning SHBs. The older workers attacked significantly more often than the younger workers. Younger honeybee workers fed the beetles significantly more often than the older bees (Pirk and Neumann 2013). The level of aggressiveness or activity might influence the outcome of the interactions between honeybees and SHBs. If the beetles are ignored or seen as not being a threat to the colony and fed by their hosts, the colony may fall victim to the beetles. If, however, there are enough bees to imprison the beetles or to remove them from the hive, this will ensure that the parasite is kept under control and diminishes the chances of the colony absconding due to SHB pressure.

Therefore, it seems that there are only quantitative and no qualitative differences between the subspecies and colony activity levels may play a crucial role. We are only beginning to understand the host/parasite interactions between honeybees and SHBs, and further research is required to fully understand this threatening, yet fascinating, host/parasite system.

The rule of thumb is to help colonies to help themselves by maintaining strong colonies and ensuring that workers have access to all parts of the hive where beetles hide and/or reproduce. Preventative measures should also be put in place so SHB are unable to reproduce outside the hive, in the honeyroom or in old, stored comb.

As one can see from the life cycle (Fig. 7) and the biology of the beetle, there are several time periods at which populations of SHB are susceptible to control strategies, where one can try to reduce and perhaps even eliminate SHBs from an apiary.

1. Make it as difficult as possible for SHB to enter the hive. Ensure that the colony is strong and there are enough guards to patrol the entrances. Decreasing the size of the entrances will also help deter infiltration by SHB (Ellis et al. 2002a, 2003a; Neumann et al. 2013); the bees patrolling the entrance will have less territory to cover, thereby reducing the probability of a SHB to successfully enter the hive. African bees do this themselves using propolis (Fig. 6).

2. Reduce possible hiding spots in the hive. Close off and fill in any holes or cracks in the hive and lid through which SHBs may enter and/or hide from the workers. This will reduce the number of possible hiding and breeding areas.

3. Do not provide SHBs the opportunity to lay eggs. Ensure workers are able to access all areas of the hive. Frames positioned too close together prevent workers from patrolling these areas, which are prime spots for SHBs to safely lay their eggs and reproduce. By ensuring that there is proper spacing between the frames allows the workers the opportunity to clean and to maintain high levels of hygiene within the hive. One should ensure that the colony is strong and there are enough workers to patrol the hive. When worker populations decrease and the brood nest diminishes in size, over winter or a dry season, it is advisable to either adjust the space of the hive to the colony size or supply additional brood/workers to keep the SHB numbers in check (Fig. 5).

4. In cases of minor SHB infestation, take the appropriate measures (above) and leave the workers to contain the numbers. However, in the case of major SHB infiltration/reproduction, the beekeeper should intervene and remove all infected parts of the comb from the hive.

5. To avoid reinfestation, use control methods aimed at the pupae developing in the soil. Pupation takes place outside the hive, and soil treatments should be applied to help prevent reinfestation of the colonies (Lundie 1940; Neumann and Elzen 2004; Neumann and Ellis 2008). In a study by Pettis and Shimanuki (2000), 80% of SHB pupae were found in the top 10 cm of soil; 83% of which was within a 30 cm radius around the entrance to the hive. Wandering larvae are able to cover distances of 2 m or more.

Acknowledgements The author is grateful to Colleen Hepburn for commenting on an earlier version of the chapter and to Chamanti Laing and Ursula Strauss for providing photo material.

References

Allsopp MH, Hepburn HR (1997) Swarming, supersedure and mating system of a natural population of honey bees (*Apis mellifera capensis*). J Apicult Res 1:41–48

Anderson DL, Trueman JWH (2000) Varroa jacobsoni (Acari: Varroidae) is more than one species. Exp Appl Acarol 24:165–189

Bernier M, Fournier V, Eccles L, Giovenazzo P (2015) Control of *Aethina tumida* (Coleoptera: Nitidulidae) using in-hive traps. Can Entomol 147:97–108. doi:10.4039/tce.2014.28

Dietemann V, Pirk CWW, Crewe RM (2009) Is there a need for conservation of honeybees in Africa? Apidologie 40:285–295

Ellis JD, Delaplane KS, Hepburn R, Elzen PJ (2002a) Controlling small hive beetles (*Aethina tumida* Murray) in honey bee (*Apis mellifera*) colonies using a modified hive entrance. Am Bee J 142:288–290

Ellis JD, Pirk CWW, Hepburn HR, Kastberger G, Elzen PJ (2002b) Small hive beetles survive in honeybee prisons by behavioural mimicry. Naturwissenschaften 89:326–328

Ellis JD, Delaplane KS, Hepburn R, Elzen PJ (2003a) Efficacy of modified hive entrances and a bottom screen device for controlling *Aethina tumida* (Coleoptera: Nitidulidae) infestations in *Apis mellifera* (Hymenoptera: Apidae) colonies. J Econ Entomol 96:1647–1652. doi:10.1603/0022-0493-96.6.1647

Ellis JD, Hepburn HR, Ellis AM, Elzen PJ (2003b) Prison construction and guarding behaviour by European honeybees is dependent on inmate small hive beetle density. Naturwissenschaften 90:382–384. doi:10.1007/S00114-003-0447-Y

Ellis JD, Hepburn HR, Ellis AM, Elzen PJ (2003c) Social encapsulation of the small hive beetle (*Aethina tumida* Murray) by European honeybees (*Apis mellifera* L.). Insectes Soc 50:286–291. doi:10.1007/S00040-003-0671-7

Ellis JD, Hepburn R, Delaplane KS, Elzen PJ (2003d) A scientific note on small hive beetle (*Aethina tumida*) oviposition and behaviour during European (*Apis mellifera*) honey bee clustering and absconding events. J Apicult Res 42:47–48

Ellis JD, Hepburn R, Delaplane KS, Neumann P, Elzen PJ (2003e) The effects of adult small hive beetles, *Aethina tumida* (Coleoptera: Nitidulidae), on nests and flight activity of Cape and European honey bees (*Apis mellifera*). Apidologie 34:399–408. doi:10.1051/apido:2003038

Elzen PJ, Baxter JR, Neumann P, Solbrig A, Pirk C et al (2001) Behaviour of African and European subspecies of *Apis mellifera* toward the small hive beetle, *Aethina tumida*. J Apicult Res 40:40–41

Hepburn HR (1988) Absconding in the African honeybee—the queen, engorgement and wax secretion. J Apicult Res 27:95–102

Hepburn HR, Radloff SE (1998) Honeybees of Africa. Springer Verlag Berlin, Germany

Johannsmeier MF (2001) Beekeeping in South Africa. ARC-Plant Protection Research Institute

Lundie AE (1940) The small hive beetle, *Aethina tumida*. Sci B U S Afr 220:5–19

Mutinelli F, Montarsi F, Federico G, Granato A, Ponti AM et al (2014) Detection of *Aethina tumida* Murray (Coleoptera: Nitidulidae.) in Italy: outbreaks and early reaction measures. J Apicult Res 53:569–575. doi:10.3896/Ibra.1.53.5.13

Neumann P, Ellis JD (2008) The small hive beetle (*Aethina tumida* Murray, Coleoptera: Nitidulidae): distribution, biology and control of an invasive species. J Apicult Res 47:181–183. doi:10.3827/Ibra.1.47.3.01

Neumann P, Elzen PJ (2004) The biology of the small hive beetle (*Aethina tumida*, Coleoptera: Nitidulidae): gaps in our knowledge of an invasive species. Apidologie 35:229–247

Neumann P, Härtel S (2004) Removal of small hive beetle (*Aethina tumida*) eggs and larvae by African honeybee colonies (*Apis mellifera scutellata*). Apidologie 35:31–36

Neumann P, Pirk CWW, Hepburn HR, Solbrig AJ, Ratnieks FLW et al (2001) Social encapsulation of beetle parasites by Cape honeybee colonies (*Apis mellifera capensis* Esch.). Naturwissenschaften 88:214–216. doi:10.1007/s001140100224

Neumann P, Evans JD, Pettis JS et al (2013) Standard methods for small hive beetle research. J Apicult Res 52. doi:10.3896/IBRA.1.52.4.19

Neumann P, Naef J, Crailsheim K, Crewe RM, Pirk CWW (2015) Hit-and-run trophallaxis of small hive beetles. Ecol Evol 5:5478–5486. doi:10.1002/ece3.1806

Neumann P, Pettis JS, Schäfer MO (2016) Quo vadis Aethina tumida? biology and control of small hive beetles. Apidologie 1–40. doi:10.1007/s13592-016-0426-x; 10.1007/s13592-016-0426-x

Pettis JS, Shimanuki H (2000) Observations on the small hive beetle, Aethina tumida Murray, in the United States. Am Bee J 140:152–155

Pirk CWW, Neumann P (2013) Small Hive Beetles are Facultative Predators of Adult Honey Bees. Journal of Insect Behavior 26:796–803. doi:10.1007/s10905-013-9392-6

Pirk CWW, Human H, Crewe RM, vanEngelsdorp D (2014) A survey of managed honey bee colony losses in the Republic of South Africa—2009 to 2011. J Apicult Res 53:35–42. doi:10.3896/IBRA.1.53.1.03

Pirk CWW, Strauss U, Yusuf A, et al (2015) Honeybee health in Africa—a review. Apidologie: 1–25. doi:10.1007/s13592-015-0406-6; 10.1007/s13592-015-0406-6

Schäfer MO, Pettis JS, Ritter WG, Neumann P (2008) A scientific note on quantitative diagnosis of small hive beetles, Aethina tumida, in the field. Apidologie 39:564–565. doi:10.1051/apido:2008038

Schmolke MD (1974) A study of *Aethina tumida*: The small hive beetle. University of Rhodesia (Zimbabwe) Harare

Spiewok S, Neumann P (2006) The impact of recent queenloss and colony phenotype on the removal of small hive beetle (Aethina tumida Murray) eggs and larvae by African honeybee colonies (Apis mellifera capensis esch.). J Insect Behav 19:601–611. doi:10.1007/S10905-006-9046-Z

Spiewok S, Neumann P, Hepburn HR (2006) Preparation for disturbance-induced absconding of Cape honeybee colonies (*Apis mellifera capensis* Esch.). Insectes Soc 53:27–31

Strauss U, Human H, Gauthier L, Crewe RM, Dietemann V et al (2013) Seasonal prevalence of pathogens and parasites in the savannah honeybee (*Apis mellifera scutellata*). J Invertebr Pathol 114:45–52. doi:10.1016/j.jip.2013.05.003

Torto B, Boucias DG, Arbogast RT, Tumlinson JH, Teal PEA (2007) Multitrophic interaction facilitates parasite-host relationship between an invasive beetle and the honey bee. Proc Natl Acad Sci USA 104:8374–8378

Villa JD (2004) Swarming behavior of honey bees (Hymenoptera: Apidae) in Southeastern Louisiana. Ann Entomol Soc Am 97:111–116

Author Biography

Christian W.W. Pirk is a Professor in the Department of Zoology and Entomology at the University of Pretoria and a member of the Academy of Science of South Africa. Christian did his Ph.D. from 2000 to 2002 under the supervision of Prof. R. Hepburn at Rhodes University (Grahamstown, South Africa) on the topic of "reproductive conflicts in honeybees."

Thereafter, he was a Postdoctoral Fellow in Prof. Tautz's group at the University of Würzburg followed by joining Prof. Moritz's group at Halle University. In 2005 he joined Prof. Crewe's laboratory at the University of Pretoria, and in 2009, he accepted a faculty position in the Department of Zoology and Entomology; two years later, he was promoted to associate professor and he has been a full professor since 2015. His main research focus is on social insects, using a multidisciplinary approach by combing mathematics, chemistry, behavioral studies, population analysis, and molecular ecology. Moreover, researching self-organisation in social insects, the organisation of groups, mechanisms of coordination and task allocation and the role and means of communication in achieving coherent collective behavior, and its applications in industrial processes. Another field of interest is the interaction and coevolution of hosts and parasites/pathogens, for example, those between the honeybee and small hive beetle/brood diseases. Christian leads the Social Insects Research Group, which is a vibrant group of more than 20 members including faculty members, postdocs, and postgraduates. To date, he has published over 100 peer-reviewed articles, authored the book "Honeybee Nests," contributed to five chapters, including three chapters to the book "Honeybees of Asia."

Foulbrood Diseases of Honey Bees— From Science to Practice

Elke Genersch

Abstract

Bacterial diseases are quite common in animals, and in most cases one animal is host to many bacterial pathogens. However, only two bacteria are known to cause disease in honey bee and these two bacteria are only pathogenic to honey bee larvae. One of these bacterial brood pathogens is the bacterium *Paenibacillus larvae* (*P. larvae*), causing American Foulbrood (AFB). AFB is the most serious honey bee brood disease because AFB is not only able to kill infected larvae but may also lead to the death of entire colonies. AFB is highly contagious and can spread quite fast within an apiary and between apiaries. Therefore, AFB is a notifiable epizootic in many countries and mandatory control measures often include the culling of diseased colonies or even of all colonies of an affected apiary. So far, no satisfactory strategies which can prevent or cure the disease are available. To change this situation, a better understanding of the pathogenesis of AFB is urgently needed. In the following chapter, the current state of AFB research is presented. The two *P. larvae* genotypes ERIC I and ERIC II are introduced, and their relevance for both research and practice is explained. Several *P. larvae* factors necessary for pathogenesis and virulence are described together with the methods used for their discovery and evaluation. A model on the interaction between *P. larvae* and larvae during pathogenesis is presented integrating all the novel data. Special emphasis has been put into outlining the practical implications of the newly generated knowledge.

E. Genersch (✉)
Institute for Bee Research, Friedrich-Engels-Str. 32, 16540 Hohen Neuendorf, Germany
e-mail: elke.genersch@hu-berlin.de; elke.genersch@fu-berlin.de

© Springer International Publishing AG 2017
R.H. Vreeland and D. Sammataro (eds.), *Beekeeping – From Science to Practice*,
DOI 10.1007/978-3-319-60637-8_10

1 Introduction

Foulbrood diseases of honey bees have a long history. A description of a honey bee disease condition which "is indicated in a lassitude on the part of the bees and in a malodorous hive" can already be found in the History of Animals (*Historia animalium* IX.40), written by Aristotle more than 2000 years ago (Aristotle 350 B.C.). The name foulbrood was coined for such diseases in 1766 by the Sorbian-German clergyman and naturalist Adam Gottlob Schirach, who had already described two different conditions characterized by fouling larvae, the "real" and the "false" foulbrood pest (Schirach 1766). This differentiation already gave evidence for the existence of honey bee diseases resembling what we know as European Foulbrood and American Foulbrood. Schirach had no clue about the infectious etiology of the disease, although at that time microorganisms and bacteria had already been described by Anthony Leewenhoeck as "*animalcula* or living atoms" (van Leewenhoeck 1677–1678) or rather "small living animals, which moved themselves very extravagantly" (van Leewenhoeck 1684). However, nobody was aware of the possible relation between bacteria and diseases until in the second half of the nineteenth century, Louis Pasteur and Robert Koch came up with the "germ theory" introducing microorganisms as causative agents of fermentation and different infectious diseases (Koch 1878, 1893; Pasteur 1866, 1909–1914). Shortly thereafter, in 1885, Cheshire and Cheney isolated *Bacillus alvei* from foulbrood diseased hives and concluded that *B. alvei* is the causative agent of foulbrood of honey bees (Cheshire and Cheney 1885). This view had to be partially revised, when in 1906 the American scientist White failed to isolate *B. alvei* from the remains of larvae that died from foulbrood and instead consistently isolated a pure culture of a new bacterium which he called *Bacillus larvae* (White 1906). White realized that actually two different foulbrood diseases existed, one caused by *Bacillus larvae* and the other one involving *B. alvei*. In acknowledging the pioneering work of Cheshire and Cheney, White proposed calling the latter one European Foulbrood (EFB) while the foulbrood disease caused by *B. larvae* should accordingly be named American Foulbrood (AFB). It soon became clear that American Foulbrood was the "real foulbrood pest" being much more serious and devastating than the rather being European Foulbrood, the "false foulbrood pest."

Due to its relevance for the beekeeping sector, AFB attracted considerable scientific attention. Over the last hundred years, many researchers, most of them being bee scientists, investigated different aspects of AFB and its etiological agent, a bacterium now known as *Paenibacillus larvae* (Genersch et al. 2006). When the disease AFB was the focus of research, experimental work was usually performed with colonies, and transmission between colonies and infection of colonies was studied. When the bacterium *P. larvae* was the focus, experiments were conducted with cultured bacteria and concentrated on culture conditions, growth inhibition, and genotyping.

From these studies, we know that the spores of *P. larvae* are the infectious form of this pathogen and that larvae are susceptible to infection only during the first 36 h after egg hatching. Exposure to only a few spores is sufficient to infect an individual larva which will eventually die from the disease. The life of the vegetative *P. larvae* bacteria does not end with the life of the larva. Instead, the larval cadaver is decomposed by the still proliferating *P. larvae* bacteria to a ropy mass. It is not before all nutrients are used up that the bacteria sporulate and the ropy mass dries down to a hard scale (AFB scale) containing billions of spores. Hence, vegetative *P. larvae* have two life phases, one as a pathogen which is killing honey bee larvae, the other as a saprophyte decomposing dead honey bee larvae. Ropy mass and AFB scales are the clinical symptoms of an AFB-diseased colony [for recent reviews: (Ashiralieva and Genersch 2006; Genersch 2007, 2008, 2010; Poppinga and Genersch 2015)]. Death of infected honey bee larvae and total degradation of the larval cadavers are the prerequisites for transmission and spreading of the disease within and between colonies. Hence, *P. larvae* can be considered an obligate killer because killing of infected larvae must be its final goal to ensure long-term survival of its species.

Despite these valuable data, experimental work aiming at understanding what is going on in infected individual larvae was lacking. Until 15 years ago, there was no scientifically sound information available on the interactions between the pathogen *P. larvae* and its host, honey bee larvae, at the cellular or molecular level. This situation was clearly unsatisfactory because for finding a cure for or preventative measures against any disease, it is essential to first understand the disease process not only in the affected population (honey bee colony in this case) but also and even more importantly in the diseased individual (honey bee larva).

2 Laboratory Infection Assays with Individual Larvae

In order to analyze the infection process in infected larvae in more detail, the first steps required were (i) the development of a method for rearing honey bee larvae in the laboratory (Crailsheim et al. 2013; Genersch et al. 2005, 2006) and (ii) the development of a laboratory infection assay (Crailsheim et al. 2013; Genersch et al. 2005, 2006) (Fig. 1). Only if the infection of honey bee larvae with *P. larvae* can be performed in the laboratory and only if subsequently the infected bee larvae are allowed to complete development in an incubator so that they can be controlled on a daily basis, will it be possible to observe all steps of the infection process in infected bee larvae. The already existing data on *P. larvae* infections (Bailey and Leed 1962; Dingman and Stahly 1983; Hoage and Rothenbuhler 1966; Hornitzky 1998; Peng et al. 1996; Tarr 1937, 1938; Woodrow 1942) were the basis for successfully establishing laboratory exposure bioassays with honey bee larvae and the bacterium *P. larvae*.

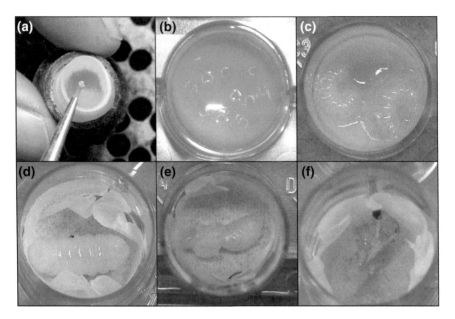

Fig. 1 *Development of healthy honey bee larvae into brown-eyed pupae in the laboratory.* Honey bee larvae are removed from the brood cells 12 h after egg hatch (**a**) and transferred in groups of ten into wells of plastic cell culture plates (24-well plates) which contain larval diet (**b**). Larvae are transferred into wells containing fresh larval diet every day (**c**). As the larvae grow in size, the number of larvae per well is reduced (**c**). With the beginning of metamorphosis, coiled/stretched larvae are placed separately into wells lined with filter paper (**d**) and pupal development is allowed to proceed (**e–f**)

It is long since known that honey bee larvae are most susceptible to infection only during the first 36 h after egg hatching (Woodrow 1942). Therefore, for infection experiments young honey bee larvae at the age of approximately 12 h after egg hatching were collected (Fig. 1a). Groups of ten honey bee larvae were placed into wells of 24-well plates filled with artificial larval diet (Fig. 1b) prepared from royal jelly, sucrose, and fructose (Genersch et al. 2005). For infection with *P. larvae*, honey bee larvae were placed onto larval food containing defined concentrations of *P. larvae* spores (Fig. 1b) and exposed to the infectious diet for 24 h. During this time, honey bee larvae will ingest the contaminated food thereby taking up *P. larvae* spores. This mimics the normal route of transmission of *P. larvae* spores to young honey bee larvae in a bee colony. After 24 h, hence at the age of 36 h after egg hatching, honey bee larvae were no longer considered susceptible and were transferred to new wells containing uncontaminated larval diet. Non-infected control larvae were fed with uncontaminated diet throughout the entire experiment. All honey bee larvae were monitored daily, and their health status was evaluated and recorded (Fig. 1c–f). Larvae were classified as dead from AFB when vegetative *P. larvae* bacteria could be isolated from honey bee larval remains via overnight cultivation on agar plates. If this was not possible, the honey bee larvae were classified as having died from manipulation or unknown reasons.

To correctly evaluate this type of infection assay, it is important to always differentiate between AFB-dead honey bee larvae and larvae that died from other causes in the infected groups; only AFB-dead honey bee larvae should be included in the calculations of total and cumulative mortality due to AFB. The assays should be performed with *P. larvae* spores originating from one apiary or AFB outbreak only, and *P. larvae* spores from different sources must not be mixed. This is necessary in order to ensure that defined *P. larvae* strains are used for infection. These strains can then be further characterized in the laboratory to gain a better understanding of the infection process (Gensersch et al. 2006; Gensersch and Otten 2003; Neuendorf et al. 2004).

The experimental groups of honey bee larvae always consisted of a mixture of larvae originating from different honey bee colonies. This was to minimize the effect of the bees' genetic background on the outcome of the infection and to be able to observe the effect of the pathogen on the host as best as possible (Crailsheim et al. 2013). Ideally, the replicates of the experiments were performed at different time points during the bee season to discount if seasonal or weather effects could influence the results (Crailsheim et al. 2013).

In honey bee colonies, natural mortality of worker larvae has been reported to be approximately 15% (Fukuda and Sakagami 1968). Hence, any assays with higher mortality in the non-infected control groups should be considered invalid (Crailsheim et al. 2013). This threshold represents a quality control for the infection experiments because AFB lethality can only be analyzed if few larvae died from manipulation: A honey bee larva can only die once, either from manipulation or from infection. The more honey bee larvae die from manipulation, the fewer larvae can die from experimentally induced AFB. However, early in the bee season as well as late in the bee season and during bad weather periods, control mortality exceeding 15% could be observed. Obviously, sometimes honey bee larvae are more vulnerable than normal and, hence, in these cases mortality rates in the control groups of up to 20% can be accepted (Crailsheim et al. 2013). Likewise, if in the infection groups more than 15% of the honey bee larvae died from other causes than *P. larvae* infection, these experiments must also be considered invalid. By following all these rules and precautionary measures, laboratory infection assays are a perfect means to analyze how *P. larvae* kills honey bee larvae.

3   What Can We Learn from Laboratory Infection Assays with Individual Larvae?

Until about a decade ago, nearly all text books on AFB stated that AFB-diseased honey bee larvae die in the capped brood cells and hence, during pupal development. This perception of AFB originated from bee biologists who in former times were mostly interested in the honey bee colony as host and, hence, studied AFB predominantly in the context of the honey bee colony. This resulted in false ideas about how the disease is acting at the level of the individual larva. The statement that AFB-diseased larvae die in the capped stage was based on the fact that in diseased honey bee colonies, only those larvae can be observed by bee keepers, veterinarians, and scientists which survive until cell capping and die in the capped stage; those larvae that were removed by the nurse bees were kind of invisible to the scientists and others.

However, laboratory infection experiments, performed as described in the previous section, revealed that the first infected larvae had already died at 2 or 3 days post-infection (Fig. 2), hence, at the age of 3 or 4 days after egg hatching and, therefore, clearly before capping of the brood cell, which coincides with the onset

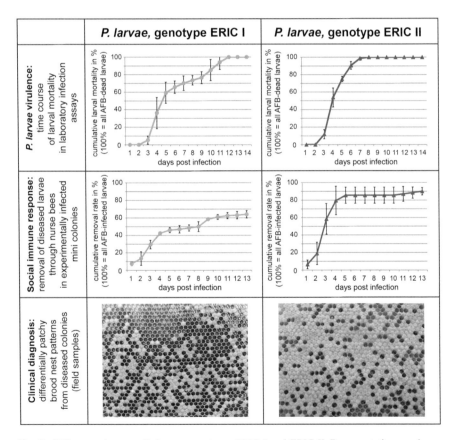

Fig. 2 *Differences between P. larvae genotypes ERIC I and ERIC II.* Representative graphs are shown to illustrate the differences between the two genotypes in respect of larval mortality as determined in laboratory infection assays (*upper row*) and in respect of the removal of diseased larvae determined in experimentally infected mini-colonies (*middle row*). To illustrate how these differences can influence the clinical picture of AFB, representative pictures of the appearance of the brood nest in diseased colonies in the field are shown (*lower row*)

of metamorphosis (Genersch et al. 2005, 2006). Therefore, larval mortality due to *P. larvae* infection is not restricted to a certain time window but rather occurs throughout the entire larval and pupal development following infection (Fig. 2). A closer look at the curves shown in the upper row of Fig. 2 reveals that actually the proportion of larvae that are dying before cell capping (around day 6 post-infection) represents the majority of AFB-dead larvae. These larvae will be removed by nurse bees as part of the hygienic behavior (Fig. 2, middle row). Hence, in former times the major part of the disease was scientifically neglected because it was not obvious at the colony level. The scientists, bee keepers, veterinarians just couldn't and didn't see these larvae and concluded that they did not exist. The consequences for the classical clinical diagnosis based on detecting AFB-dead larvae in capped cells are obvious: Such diagnoses will always only detect very late stages of AFB disease in the honey bee colony, that is, when the proportion of larvae that could not be cleansed out becomes obvious when looking at a comb.

In contrast to observing infected and diseased honey bee colonies in the field, performing exposure bioassays with individual larvae in the laboratory allowed collecting data on daily larval mortality and, based on these data, to generate the mortality curves in relation to the times which are shown in Fig. 2. Many different *P. larvae* strains isolated from different AFB outbreaks were analyzed like this in order to detect common patterns and differences. And indeed, a detailed comparison of the obtained mortality curves revealed that they fell into two groups, each having a characteristic pattern: The mortality curves either displayed a biphasic course with two exponential phases of mortality in the early and late phases of infection, separated by a phase of reduced mortality between day 5 and day 9 post-infection or were sigmoidal in shape with 90% of the infected larvae being dead by day 6 post-infection (Fig. 2). In the first case, a considerable proportion of larvae (approximately 40% on average) survived until the beginning of metamorphosis (at the time of cell capping in the colony) and died between day 9 and 13 post-infection (capped cell stage in the colony) (Fig. 2, Table 1), while in the latter case nearly all larvae died before cell capping and only few larvae survived until metamorphosis (Fig. 2, Table 1) (Genersch et al. 2005). Therefore, the second important result from the laboratory infection assays was the discovery that the *P. larvae* strains isolated from current AFB outbreaks form two groups, which differ in virulence at the level of the individual larva. This difference can best be described as difference in the time course of larval mortality: Either diseased larvae die predominantly already during larval development and, hence, in the open cell, or a considerable proportion of the diseased larvae survive until the beginning of the pupal development, then cease to develop, and die as stretched larva in the capped cell. The relevance of this difference will be explained below.

The two groups of *P. larvae* strains differing in virulence as determined by exposure bioassays correlated with two genotypes of *P. larvae*, *P. larvae* ERIC I and ERIC II (Fig. 2, Table 1), which had been defined by molecular methods at around the same time (Genersch et al. 2006). These *P. larvae* genotypes are causing contemporary AFB outbreaks worldwide (Morrissey et al. 2015), and, hence, the

Table 1 Differences between *P. larvae* ERIC I and ERIC II

		P. larvae, genotype ERIC I	*P. larvae*, genotype ERIC II	References
P. larvae virulence parameters determined in laboratory infection assays *	LT_{100} (time it takes *P. larvae* to kill all infected larvae)	~13 days	~7 days	Genersch et al. (2005, 2006)
	proportion of infected larvae dying before cell capping	~40–60%	~80–95%	Genersch et al. (2005, 2006)
	proportion of infected larvae dying in capped cells and developing into ropy mass and foulbrood scales	~60–40%	~20–5%	Genersch et al. (2005) Rauch et al. (2009)
Experimentally proven virulence factors of *P. larvae* **	Chitin-degrading enzyme *Pl*CBP49	+	+	Garcia-Gonzalez et al. (2014c)
	Toxin Plx1	+	–	Fünfhaus et al. (2013)
	Toxin Plx2	+	–	Fünfhaus et al. (2013)
	S-layer protein SplA	–	+	Poppinga et al. (2012)
	Antibiotic paenilamicin	–	+	Garcia-Gonzalez et al. (2014b) Müller et al. (2014)

Note *Experimental infection of honey bee larvae and honey bee mini-colonies in the laboratory revealed differences in virulence between the two genotypes, *P. larvae* ERIC I and ERIC II. **Molecular analyses followed by functional assays resulted in the identification of virulence factors of *P. larvae* and in the determination of their role during pathogenesis

question arose whether these differences are also relevant for the field and the veterinary and beekeeping practice.

4 Relevance of Laboratory Findings for the Field and Practice

The above-described genotype-specific differences in the time course of larval mortality at the individual larval level led to the hypothesis that these differences will also have an impact on *P. larvae* virulence at the colony level in the field. In the laboratory, most larvae infected with *P. larvae* ERIC II died before they reached metamorphosis, while a considerable proportion of larvae infected with *P. larvae*

ERIC I died in the laboratory after the onset of metamorphosis (Fig. 2). To prove the relevance of these laboratory findings for the beekeeping and veterinary practice, experimental infection of larvae in mini-colonies in the laboratory—of course under extreme safety conditions—was performed. Two hypotheses needed to be analyzed: (i) In queen-right mini-colonies, nurse bees engaged in hygienic behavior will remove most, if not all, of the infected larvae that died before metamorphosis (before their brood cells would be capped); (ii) this hygienic behavior will result in fewer capped cells containing degraded larvae or the ropy mass in the case of an infection with *P. larvae* ERIC II than after infection with *P. larvae* ERIC I.

For controlled infection experiments with larvae in queen-right mini-colonies, the queens were caged so that they laid their eggs in defined areas of the brood combs. Twelve hours after larvae had hatched from these eggs, they were individually fed a drop of artificial larval diet containing a defined dose of *P. larvae* spores (infected groups) or not containing any spores (control groups). For each tested strain, a dose sufficient to yield the 100% lethal dose was used. Therefore, the exact dose of spores varied from strain to strain. However, in most cases a dose between 10 and 50 spores per larva was sufficient to successfully infect and kill all exposed larvae. Manipulated cells were marked on a see-through plastic sheet put on the comb. This allowed following the fate of each infected larva during the daily inspections. Cleansed out cells indicative for the removal of larvae through nurses engaged in hygienic behavior were recorded, and the time course of larval removal was depicted graphically (Fig. 2, *middle row*). After 13 days, all capped cells still containing the originally infected larvae were opened and the health status of the larvae was examined.

The results confirmed the hypothesis: Much more AFB-dead larvae in capped cells were found in the *P. larvae* ERIC I—than in the *P. larvae* ERIC II-infected comb areas (Rauch et al. 2009). In the latter case, most infected larvae had been removed by nurse bees (Rauch et al. 2009). The proportions of dead larvae found under the cappings corresponded to the numbers determined in the laboratory for the proportion of larvae that died after the onset of metamorphosis (Fig. 2, *upper row*). In colonies, the ropy mass in capped cells will dry down to the so-called foulbrood scales containing billions of spores which facilitate the disease spreading throughout the colony. The more larvae that die after being capped over and without being removed by nurse bees, the more foulbrood scales will develop, and the more spores will circulate in the colony. Provided that the same number of larvae are infected and die, more newly generated infectious spores will be produced in ERIC I-infected colonies than in ERIC II-infected colonies (Table 1). Hence, disease development and resulting colony collapse will be much faster in ERIC I-infected colonies than in ERIC II-infected colonies, and *P. larvae* ERIC I must be considered more virulent at colony level than *P. larvae* ERIC II. The reverse virulence at individual and colony level is counterintuitive, but it is very important to understand this relation: The fast killing and, hence, more virulent genotype at the larval level becomes the less virulent genotype at the colony level, because the bees' social immune response can cope much better with it by effectively removing most (but not all!) of the infected larvae, thereby reducing the pathogen levels in the colony. In contrast, the bees' social immune response less efficiently wards off the genotype *P. larvae* ERIC I that kills larvae much slower and, hence, is less virulent at the larval level; larvae that die too late (i.e., after cell capping) are rather not removed from the colony and instead converted by the pathogen into bacterial biomass and eventually into bacterial spores. Nevertheless, since both genotypes are lethal for individual larvae, they are also lethal for entire colonies and cause AFB outbreaks. However, due to the differences in the numbers of diseased larvae that are not removed by nurse bees and remain in their cells until the final stage of disease (Table 1, Fig. 2), infections with *P. larvae* ERIC II normally develop much more slowly in the colony and clinical symptoms may be evident only after several years of "subclinical" disease. In contrast, colonies infected by *P. larvae* ERIC I develop clinical symptoms much faster and may become obtrusive within a couple of months after infection.

The described (genotype-specific) differences in virulence at the colony level are important for understanding *P. larvae*, AFB, and disease progression within colonies. However, another aspect of these results is of even greater importance for beekeeping and veterinary practices, and that is the impact these results have on the clinical diagnosis of diseased colonies. The ropy mass in capped cells and foul-brood scales are **the** clinical symptoms used to diagnose AFB in the field. If colonies are weak and have a patchy broodnest, the beekeeper and veterinarian might suspect AFB. However, they will have little chance to correctly diagnose AFB by visually inspecting a colony if the specific AFB symptoms are difficult to spot—as is the case for *P. larvae* ERIC II infections (Fig. 2, *lower row*). Remember, the proportion of AFB-dead larvae that will have died in the capped stage will only be around 10% for ERIC II-infected colonies (Table 1). In contrast, in ERIC I-infected colonies the same number of AFB-dead larvae will yield about four times more dead larvae in capped cells (Table 1). Because only the larval remains (ropy mass or scales) in capped cells are used for the clinical diagnosis of AFB, this diagnosis is four times easier in the case of ERIC I-infections than in the case of ERIC II infections. Or—in other words—compared to ERIC I-infected colonies, four times more larvae need to have died in ERIC II-infected colonies before clinical diagnosis is as easy as it is in ERIC I-infected colonies. In addition, cells cleansed out by nurse bees will be reused for egg laying by the queen. The more cells are cleansed out and again used for larval rearing in the colony, the less patchy the brood nest and the less obvious an ongoing AFB disease will be (Fig. 2, lower row).

In the absence of a correct clinical diagnosis, suitable control measures will not be initiated and the disease will spread unhampered throughout an apiary or region. It is important to emphasize here that both *P. larvae* genotypes infect and kill larvae and that both are virulent and dangerous for infected colonies. The social immune response of the honey bees toward brood diseases can slow down disease progression in colonies infected with *P. larvae* ERIC II but less so in colonies infected with *P. larvae* ERIC I. In both cases, colonies will eventually succumb to the disease if left untreated and will be a source of infectious spores for neighboring colonies.

In Germany, AFB is a notifiable disease and burning of diseased colonies is widely recommended or even stipulated by the authorities. Therefore, in several Federal States of Germany, AFB monitoring programs are in place and outbreaks are recorded along with their history and follow-up records. Generally, two categories of outbreaks stand out:

(i) Outbreaks that are difficult to diagnose in the field because only a few, hardly detectable cells containing decomposed larvae (ropy mass) can be found by visual inspection, despite high levels of *P. larvae* spores present in the brood comb honeys;

(ii) Outbreaks that are easily diagnosed in the field because they are accompanied by the classical symptoms, i.e., patchy brood nest with capped cells containing larval remains forming a ropy mass or cells with tightly adhering foulbrood scales.

Genotyping the causative strains of these outbreaks revealed in the majority of cases that *P. larvae* ERIC II caused the difficult-to-diagnose outbreaks while ERIC I-outbreaks generally did not pose diagnostic problems. These records underpin the practical relevance of the recently discovered virulence differences between the recently defined *P. larvae* genotypes ERIC I and II.

The question for the beekeepers and veterinarians now is how to react to these new insights and how to overcome the problems with clinically diagnosing *P. larvae* ERIC II infections. It is evident that the statement "AFB-diseased larvae die in the capped stage" is wrong and that relying on a routine visual inspection of colonies for spotting AFB symptoms might result in overlooking early disease stages of AFB—especially in the case of *P. larvae* ERIC II infections. A possible solution to this problem is to base the diagnosis "AFB-negative" or "AFB-positive" not only on the visual inspection of a honey bee colony but to always also perform a laboratory analysis of bees or brood comb honey for the presence of *P. larvae* spores. In *P. larvae* positive samples, the genotype of *P. larvae* should be determined. This information will then help the veterinarian or bee inspector to better evaluate the brood comb pattern or clinical picture of the colony in question (Fig. 2) thereby reducing the proportion of false negative clinical diagnoses.

5 Trying to Understand the Pathogenesis of *P. Larvae* Infections

A logical consequence of discovering the genotype-specific differences in *P. larvae* virulence was understanding these differences. Virulence of a bacterial pathogen is a phenotypic feature of this pathogen. Most phenotypes are genetically determined. Hence, pathogens which differ in virulence should also differ in their genomes or should at least harbor different sets of virulence genes. Therefore, identifying genomic differences between the *P. larvae* genotypes should help to explain the

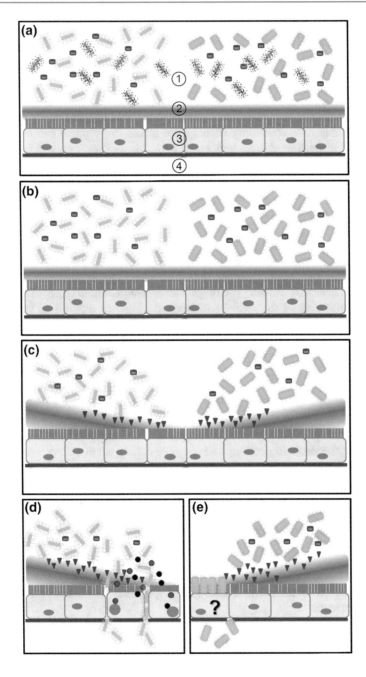

◀ **Fig. 3** *Proposed model for the molecular pathogenesis of P. larvae infections.* The proposed model integrates the virulence factors which have been proven to be involved in the virulence of *P. larvae* ERIC I and II. Bacterial and fungal competitors (*red* and *blue* bacteria) of *P. larvae* ERIC I (*green* bacteria without surface layer, *left*) and *P. larvae* ERIC II (*green* bacteria with surface layer, *right*) are eliminated via secreted secondary metabolites (*orange barrels*) (**a**). Thus, a pure culture of *P. larvae* occupies the larval gut (**b**). The chitin-degrading enzyme *Pl*CBP49 (*red triangles*), a key virulence factor of *P. larvae*, metabolizes chitin and is responsible for degrading the peritrophic matrix (**c**). The strategies used by *P. larvae* ERIC I and ERIC II for breaching the epithelial layer differ: *P. larvae* ERIC I destroys the epithelial cells via the activity of specific toxins, Plx1 and Plx2 (*blue* and *red circles* in (**d**)), while *P. larvae* ERIC II needs to adhere to the epithelial cells via its surface layer protein SplA (*dark green edge* around the bacteria) before unknown factors allow breaching the epithelial barrier (**e**). 1, midgut lumen; 2, peritrophic matrix; 3, epithelial cell layer; 4, hemocoel

observed genotype-specific phenotypic differences. The first attempts to identify virulence determinants of *P. larvae* were undertaken when only a draft genome of *P. larvae* ERIC I was published (Qin et al. 2006), but the genome data for *P. larvae* ERIC II were missing altogether. Therefore, it was necessary to use an experimental approach which is independent from preexisting genome data but is instead based on comparing the isolated bacterial genomic DNA and finding as well as sequencing genome fragments differing between the bacterial strains in question (Diatchenko et al. 1996), in this case between *P. larvae* ERIC I and ERIC II (Fünfhaus et al. 2009). Genomic regions differing between differentially virulent but otherwise closely related bacteria are likely candidate regions for virulence factors (Fünfhaus et al. 2009). In addition to comparing genomes, it is also possible to compare the expressed proteins, the so-called proteomes, of closely related bacteria to find differentially expressed proteins which might also represent putative virulence factors (Fünfhaus and Genersch 2012). Both approaches could be successfully applied to comparing *P. larvae* ERIC I and ERIC II. They revealed that *P. larvae* ERIC I genomes encode functional toxin genes which are absent from *P. larvae* ERIC II genomes (Table 1, Fig. 3) (Fünfhaus et al. 2013) and that *P. larvae* ERIC II but not *P. larvae* ERIC I expresses a functional surface layer (S-layer) protein (Table 1, Fig. 3) (Fünfhaus and Genersch 2012; Poppinga et al. 2012). Furthermore, in the genomes of both genotypes giant gene clusters were found which encode enzyme complexes responsible for the synthesis of antibiotics (Fünfhaus et al. 2009) which may be important for *P. larvae* when having to compete with the larval microbiome during infection (Table 1, Fig. 3). Whole-genome sequencing of two *P. larvae* strains of ERIC I and ERIC II followed by a detailed bioinformatic comparison of the sequence data (Djukic et al. 2014) confirmed and extended the previous results. So far, several virulence factors of *P. larvae* could be identified; some are specific for one genotype only, and others are expressed and used by both genotypes during pathogenesis (Table 1, Fig. 3).

The identification of virulence genes and proteins with a putative role in pathogenesis is interesting, but the presence of a gene or protein, no matter how interesting it is, does not prove its role or relevance in the infection process. In molecular microbiology, such proof can be provided by genetically manipulating the pathogen of interest and by creating mutants that are no longer able to express a certain factor. If infection assays reveal that these gene inactivation mutants have a reduced virulence, compared to the wild-type bacteria, it is safe to conclude that the product of the gene in question is a virulence factor.

This kind of experiments was performed for all virulence factors of *P. larvae* known so far (Table 1): The genes coding for the chitin-degrading enzyme *Pl*CBP49 (Garcia-Gonzalez et al. 2014c), the toxins Plx1 and Plx2 exclusively expressed by *P. larvae* ERIC I (Fünfhaus et al. 2013), the *P. larvae* ERIC II-specific S-layer protein SplA (Poppinga et al. 2012), and several antibiotics (Garcia-Gonzalez et al. 2014a, b; Hertlein et al. 2014, 2016; Müller et al. 2014) were individually inactivated and the virulence and lethality of all these mutant *P. larvae* strains were analyzed in laboratory infection assays in comparison with the corresponding wild-type *P. larvae* strains.

The chitin-degrading enzyme *Pl*CBP49 expressed by *P. larvae* ERIC I as well as by *P. larvae* ERIC II (Table 1) was shown to be responsible for degradation of the chitin-rich peritrophic matrix actually intended for protecting the honey bee larval gut against pathogen attack (Fig. 3). It turned out to be a key virulence factor of *P. larvae*: In the absence of *Pl*CBP49 expression, degradation of the peritrophic matrix is hampered and honey bee larval mortality is nearly abolished (Garcia-Gonzalez and Genersch 2013; Garcia-Gonzalez et al. 2014c).

The ERIC I-specific toxins Plx1 and Plx2 (Table 1) were demonstrated to act on the epithelial cells of the larval midgut and to help the bacteria breaching the epithelial barrier (Fig. 3). Lack of any of these toxins significantly reduced the ability of *P. larvae* to kill infected honey bee larvae (Fünfhaus et al. 2013).

The *P. larvae* S-layer protein SplA, only expressed by representatives of the genotype ERIC II (Table 1), was shown to play an important although not yet fully understood role in the killing strategy employed by *P. larvae* ERIC II. SplA is necessary for bacterial adhesion to the midgut epithelium (Fig. 3) because in the absence of SplA, lethality of the bacteria is significantly reduced (Poppinga et al. 2012). But so far, the steps following bacterial adhesion to epithelial cells and which are necessary for *P. larvae* ERIC II to accomplish breaching the epithelium remain elusive.

Representatives of both *P. larvae* genotypes harbor in their genomes complex giant gene clusters which encode multi-enzyme complexes responsible for the synthesis of antibiotics (Müller et al. 2015). One of these antibiotics is paenilamicin, a novel substance only synthetized by *P. larvae* ERIC II (Table 1) (Garcia-Gonzalez et al. 2014b). In laboratory infection bioassays, when honey bee larvae were infected with wild-type, paenilamicin-producing *P. larvae* bacteria or mutant *P. larvae* strains lacking paenilamicin production, presence or absence of paenilamicin production did not make a difference in the mortality of the infected honey bee larvae (Garcia-Gonzalez et al. 2014b). However, these experiments are typically performed under sterile conditions, hence in the absence of any microbiome normally presents in honey bee larvae. When bacterial competitors were added, paenilamicin was demonstrated to play a role in eliminating these bacterial competitors in the larval gut and presumably during the decomposition of the larval cadaver (Müller et al. 2014). Therefore, the discovered antibiotics might have an important role during the natural infection process (Fig. 3).

6 A Model of the Molecular Pathogenesis of *P. Larvae* Infections

The experimental results on these newly identified *P. larvae* virulence factors provided enough information to start building a draft model on how *P. larvae* accomplishes attacking, and killing honey bee larvae. Presently, the following model can be built: Vegetative *P. larvae* thriving in the larval midgut lumen secrete antibiotics which eliminate bacterial and fungal competitors (Fig. 3a) and ensure a pure culture of *P. larvae* conquering the larva (Fig. 3b). The main source of bacterial nutrition during this phase of infection is the food provided by the nurse bees and ingested by the honey bee larva. With the growing peritrophic matrix, which is actually designed to protect the larval midgut against pathogen attack, chitin becomes available to *P. larvae* as another food source. By metabolizing this chitin, *P. larvae* prevent the formation of a proper peritrophic matrix or degrade the existing peritrophic matrix structures (Fig. 3c) thus enabling the bacteria to directly attack the epithelial cells of the larval midgut. The strategies used for attacking the epithelium obviously differ between the *P. larvae* genotypes ERIC I and ERIC II as evidenced by the different genotype-specific virulence factors: *P. larvae* ERIC I attacks via specific toxins (Fig. 3d) while *P. larvae* ERIC II first adheres to the epithelial cells before unknown factors help to breach the epithelial barrier (Fig. 3e). In both cases, the infected honey bee larvae are dead in the end and are degraded to a ropy mass which dries down to a scale tightly adhering to the wall of the brood cell (AFB scale).

Although the complex interaction between *P. larvae* and its only known host, the honey bee larvae, is still not fully understood, at least the first set of virulence factors of *P. larvae* has been identified and its relevance has been proven. For sure, the quest for *P. larvae* virulence factors will be continued so that a complete picture on how *P. larvae* kills honey bee larvae will emerge in the future. Once the essential factors needed by *P. larvae* to accomplish killing and decomposing honey bee larvae are known, it will be feasible to develop sustainable control and treatment regimens for this devastating honey bee brood disease.

References

Aristotle (350 B.C.) History of Animals, Book IX, Chapter 40

Ashiralieva A, Genersch E (2006) Reclassification, genotypes, and virulence of *Paenibacillus larvae*, the etiological agent of American Foulbrood in honeybees—a review. Apidologie 37:411–420

Bailey L, Leed DC (1962) *Bacillus larvae* its cultivation *in vitro* and its growth *in vivo*. J Gen Microbiol 29:711–717

Cheshire FR, Cheyne WW (1885) The pathogenic history and history under cultivation of a new bacillus (*B. alvei*) the cause of a disease of bee hives hitherto known as foul brood. J Roy Microscop Soc 5:581–601

Crailsheim K, Brodschneider R, Aupinel P, Behrens D, Gensersch E, Vollmann J, Riessberger-Gallé U (2013) Standard methods for artificial rearing of *Apis mellifera* larvae. J Apicult Res 52:1–15

Diatchenko L, Lau Y, Campbell A, Chenchik A, Moqadam F, Hunag B, Lukyanov S, Lukyanov K, Gurskaya N, Sverdlov E, Siebert P (1996) Suppression subtractive hybridization: A method for generating differentially regulated or tissue-specific cDNA probes and libraries. Proc Natl Acad Sci USA 93:6025–6030

Dingman DW, Stahly DP (1983) Medium promoting sporulation of *Bacillus larvae* and metabolism of medium components. Appl Environ Microbiol 46:860–869

Djukic M, Brzuszkiewicz E, Fünfhaus A, Voss J, Gollnow K, Poppinga L, Liesegang H, Garcia-Gonzalez E, Gensersch E, Daniel R (2014) How to kill the honey bee larva: Genomic potential and virulence mechanisms of *Paenibacillus larvae*. PLoS ONE 9:e90914

Fukuda H, Sakagami SF (1968) Worker brood survival in honey bees. Res Popul Ecol 10:31–39

Fünfhaus A, Gensersch E (2012) Proteome analysis of *Paenibacillus larvae* reveals the existence of a putative S-layer protein. Environ Microbiol Rep 4:194–202

Fünfhaus A, Ashiralieva A, Borriss R, Gensersch E (2009) Use of suppression subtractive hybridization to identify genetic differences between differentially virulent genotypes of *Paenibacillus larvae*, the etiological agent of American Foulbrood of honeybees. Environ Microbiol Rep 1:240–250

Fünfhaus A, Poppinga L, Gensersch E (2013) Identification and characterization of two novel toxins expressed by the lethal honey bee pathogen *Paenibacillus larvae*, the causative agent of American foulbrood. Environ Microbiol 15:2951–2965

Garcia-Gonzalez E, Gensersch E (2013) Honey bee larval peritrophic matrix degradation during infection with *Paenibacillus larvae*, the aetiological agent of American foulbrood of honey bees, is a key step in pathogenesis. Environ Microbiol 15:2894–2901

Garcia-Gonzalez E, Müller S, Ensle P, Süssmuth RD, Gensersch E (2014a) Elucidation of sevadicin, a novel nonribosomal peptide secondary metabolite produced by the honey bee pathogenic bacterium *Paenibacillus larvae*. Environ Microbiol 16:1297–1309

Garcia-Gonzalez E, Müller S, Hertlein G, Heid NC, Süssmuth RD, Gensersch E (2014b) Biological effects of paenilamicin, a secondary metabolite antibiotic produced by the honey bee pathogenic bacterium *Paenibacillus larvae*. MicrobiologyOpen 3:642–656

Garcia-Gonzalez E, Poppinga L, Fünfhaus A, Hertlein G, Hedtke K, Jakubowska A, Gensersch E (2014c) *Paenibacillus larvae* chitin-degrading protein *Pl*CBP49 is a key virulence factor in American Foulbrood of honey bees. PLoS Path 10:e1004284

Gensersch E (2007) *Paenibacillus larvae* and American foulbrood in honeybees. Berl Münch Tierärztl Wschr 120:26–33

Gensersch E (2008) *Paenibacillus larvae* and American foulbrood—long since known and still surprising. J Verbr Lebensm 3:429–434

Gensersch E (2010) American Foulbrood in honeybees and its causative agent, *Paenibacillus larvae*. J Invertebr Pathol 103:S10–S19

Gensersch E, Otten C (2003) The use of repetitive element PCR fingerprinting (rep-PCR) for genetic subtyping of German field isolates of *Paenibacillus larvae* subsp. *larvae*. Apidologie 34:195–206

Gensersch E, Ashiralieva A, Fries I (2005) Strain- and genotype-specific differences in virulence of *Paenibacillus larvae* subsp. *larvae*, the causative agent of American foulbrood disease in honey bees. Appl Environ Microbiol 71:7551–7555

Gensersch E, Forsgren E, Pentikäinen J, Ashiralieva A, Rauch S, Kilwinski J, Fries I (2006) Reclassification of *Paenibacillus larvae* subsp. *pulvifaciens* and *Paenibacillus larvae* subsp. *larvae* as *Paenibacillus larvae* without subspecies differentiation. Int J Syst Evol Microbiol 56:501–511

Hertlein G, Müller S, Garcia-Gonzalez E, Poppinga L, Süssmuth R, Genersch E (2014) Production of the catechol type siderophore bacillibactin by the honey bee pathogen *Paenibacillus larvae*. PLoS ONE 9:e108272

Hertlein G, Seiffert M, Gensel S, Garcia-Gonzalez E, Ebeling J, Skobalj R, Kuthning A, Süssmuth RD, Genersch E (2016) Biological role of paenilarvins, iturin-like lipopeptide secondary metabolites produced by the honey bee pathogen *Paenibacillus larvae*. PLoS ONE 11: e0164656

Hoage TR, Rothenbuhler WC (1966) Larval honey bee response to various doses of *Bacillus larvae* spores. J Econ Entomol 59:42–45

Hornitzky M (1998) The pathogenicity of *Paenibacillus larvae* subsp. *larvae* spores and vegetative cells to honey bee (*Apis mellifera*) colonies and their susceptibility to royal jelly. J Apicult Res 37:267–271

Koch R (1878) Untersuchungen über die Aetiologie der Wundinfectionskrankheiten. F.C.W. Vogel, Leipzig

Koch R (1893) Ueber den augenblicklichen Stand der bakteriologischen Choleradiagnose. Zeitschrift für Hygiene und Infektionskrankheiten 14:319–338

Morrissey BJ, Helgason T, Poppinga L, Fünfhaus A, Genersch E, Budge GE (2015) Biogeography of *Paenibacillus larvae*, the causative agent of American foulbrood, using a new multilocus sequence typing scheme. Environ Microbiol 17:1414–1424

Müller S, Garcia-Gonzalez E, Mainz A, Hertlein G, Heid NC, Mösker E, van den Elst H, Overkleeft HS, Genersch E, Süssmuth RD (2014) Paenilamicin—structure and biosynthesis of a hybrid non-ribosomal peptide/ polyketide antibiotic from the bee pathogen *Paenibacillus larvae*. Angew Chem Int Ed Eng 53:10547–10828

Müller S, Garcia-Gonzalez E, Genersch E, Süssmuth R (2015) Involvement of secondary metabolites in the pathogenesis of the American foulbrood of honey bees caused by *Paenibacillus larvae*. Nat Prod Rep 32:765–778

Neuendorf S, Hedtke K, Tangen G, Genersch E (2004) Biochemical characterization of different genotypes of *Paenibacillus larvae* subsp. *larvae*, a honey bee bacterial pathogen. Microbiology 150:2381–2390

Pasteur L (1866) Études sur le vin: ses maladies, causes qui les provoquent, procédés nouveaux pour le conserver et pour le vieillir. A'Imprimerie Impériale, Paris

Pasteur L (1909–1914) The germ theory and its applications to medicine and surgery. P.F. Collier & Son, Louis Pasteuer (1822–1895). Scientific papers. Vol. XXXVIII, Part 7. The Harvard Classics, New York

Peng CYS, Mussen EC, Fong A, Cheng P, Wong G, Montague MA (1996) Laboratory and field studies on the effect of the antibiotic tylosin on honey bee *Apis mellifera* L. (Hymenoptera: Apidae) development and prevention of American foulbrood disease. J Invertebr Pathol 67:65–71

Poppinga L, Genersch E (2015) Molecular pathogenesis of American Foulbrood: how *Paenibacillus larvae* kills honey bee larvae. Curr Opin Insect Sci 10:29–36

Poppinga L, Janesch B, Fünfhaus A, Sekot G, Garcia-Gonzalez E, Hertlein G, Hedtke K, Schäffer C, Genersch E (2012) Identification and functional analysis of the S-layer protein SplA of *Paenibacillus larvae*, the causative agent of American Foulbrood of honey bees. PLoS Path 8: e1002716

Qin X, Evans JD, Aronstein KA, Murray KD, Weinstock GM (2006) Genome sequence of the honey bee pathogens *Paenibacillus larvae* and *Ascosphaera apis*. Insect Mol Biol 15:715–718

Rauch S, Ashiralieva A, Hedtke K, Genersch E (2009) Negative correlation between individual-insect-level virulence and colony-level virulence of *Paenibacillus larvae*, the etiological agent of American foulbrood of honeybees. Appl Environ Microbiol 75:3344–3347

Schirach AG (1766) Sächsischer Bienenvater; p. 637–641. Adam Jacob Spiekermann, Leipzig and Zittau

Tarr HLA (1937) Studies on American foulbrood of bees. I. The relative pathogenicity of vegetative cells and endospores of *Bacillus larvae* for the brood of the bee. Ann Appl Biol 24:377–384

Tarr HLA (1938) Studies on American foulbrood of bees. III. The resistance of individual larvae to inoculation with endospores of *Bacillus larvae*. Ann Appl Biol 25:807–814

van Leewenhoeck A (1677–1678) Observations by Mr. Anthony van Leewenhoeck concerning little animals by him observed in rain-well-sea and snow-water, as also in water wherein pepper had lain infused. Phil Trans 12:821–831

van Leewenhoeck A (1684) Microscopical observations about animals in the scurf of the teeth, the substance called worms in the nose, the cuticula consisting of scales. Phil Trans 14:568–574

White GF (1906) The bacteria of the apiary with special reference to bee disease. Technical Series U.S. Dept Agricult 14:1–50

Woodrow AW (1942) Susceptibility of honeybee larvae to individual inoculations with spores of *Bacillus larvae*. J Econ Entomol 35:892–895

Author Biography

Prof. Dr. Elke Genersch obtained her educational background in molecular biology, virology, biochemistry, and molecular microbiology from the University of Cologne, the Ludwig-Maximilians-Universität in Munich, the Max-Planck-Institute for Biochemistry in Martinsried, and the Freie Universität Berlin. She holds a Diploma degree in Biology and a doctoral degree in Biochemistry. She got her habilitation in the field of molecular microbiology with a thesis on American Foulbrood of Honey Bees. As a postdoc, she first pursued a solid research career in human molecular medicine but when she was offered the position of the Vice Director at the Institute for Bee Research in Hohen Neuendorf, she accepted the challenge and joined the field of bee pathology in 2001. Since February 2016, she is Professor for Molecular Microbiology at the Veterinary Faculty of the Freie Universität Berlin. Prof. Genersch's research focusses on all questions of pathogen–host interactions with bees always being the host and fungi, bacteria, and viruses being the pathogens. She is especially interested in elucidating the identity, function, and role of virulence factors because she firmly believes that understanding a pathogen is the first step in developing preventive and curative measures against the respective infectious disease.

Beekeeping and Science

Yves Le Conte

Abstract

Chemical communication is one of the most fascinating areas of social insect science among which honeybee is the most known model. More than 50 pheromonal compounds had been identified in the honeybee, most of them triggering releaser effects to the receivers, and the queen mandibular pheromone had been the first primer pheromone identified in the animal kingdom. This chapter focuses on the latest findings in this topic, particularly the discovery of brood pheromones and the worker inhibitor pheromones regulating behavioral development of the nurse workers. We describe the finding of chemicals produced by the honeybee larvae involved in their recognition by the varroa mite. This work lead to the characterization of a brood pheromone involved in the recognition of the larvae and their needs by the workers and having primer effects on hypopharyngeal gland secretions and ovary development. This blend of chemical compounds also delays the age at first foraging, so that the larvae can manipulate the workers to achieve their needs. One compound of this blend, ethyl oleate (EO), had been found to be produced by forager bees to regulate the behavioral development of younger bees. So, the same compound, EO, is produced by different members in the colonies, the larvae and the foragers, to regulate the equilibrium between nurses and foragers, thus optimizing colony development. Potential uses of those pheromonal compounds in beekeeping are discussed.

Y. Le Conte (✉)
INRA, UR 406 Abeilles et Environnement, Domaine Saint Paul, Site Agroparc,
84914 Avignon Cedex 9, France
e-mail: yves.le-conte@inra.fr

1 Discovering a Honeybee Brood Pheromone

Many pheromomal compounds had been discovered in the honeybee since the first queen pheromone was identified in the 1960s (Barbier and Lederer 1960; Butler et al. 1961) as reviewed by Keeling et al. (2004) and Slessor et al. (2005). In this chapter, we will focus on the latest findings on chemical communication in the honeybee, especially brood pheromones and worker inhibitor pheromones.

The varroa mite arrived in Europe in the 1970s and in France in 1982. Very little was known about its biology, and it became the topic of investigations by numerous laboratories in Europe. In France, in the framework of my Ph.D. thesis at the research center of Bures sur Yvette (INRA-CNRS), we were searching for chemical compounds emitted by the honeybee larvae that were involved in the attraction of the mite toward its reproduction and feeding site. The first goal was to understand the parasitic mechanism; the second was to develop a device that used those attractive compounds to trap the mite inside honeybee colonies.

After a few unfruitful years, we found that temperature was very attractive to the mite, which can discriminate a difference in temperatures of 1.1 °C in a choice apparatus (see Fig. 1) and probably is a key in communication (Le Conte and Arnold 1987); its preferred temperature was around 32–33 °C (89.6–91.4 °F), which fits with the drone brood temperature (Le Conte and Arnold 1988). Then, we decided to run our experiments at 34 °C (93.2 F), as it corresponded to the average brood temperature found in the colony's brood nest. To try to find an attractive compound originating from the bee larvae, we used a four-arm olfactometer (Fig. 2). This device enables us to deliver 4 different airflows, one from each corner, to the center of the box. Each airflow can be odorized, for example, with volatile compounds from bee larvae, or extracts or chemical compounds. The behavior of mites, introduced to the center of the apparatus, can be observed in relation to the different odors in the airflows and compared with unscented ones. We found that the larval odors were attractive to the mite; then, extracts of larvae in hexane (a solvent) were also attractive: The compounds were there! We then fractionated the active extract into four different parts and found that only one fraction was attractive to the mite. This fraction contained 10 different chemical compounds, fatty acid esters, ethyl or methyl palmitate, stearate, oleate, linoleate or linolenate. We tested these individually in the olfactometer, and three of them attracted the mites (Le Conte et al. 1989).

A patent was filed for the use of those chemical compounds for varroa control. The idea was to use those attractants to trap the mite inside the hive. Different trap devices were set up, including the esters at different concentrations, and introduced in the hive. Unfortunately, some mites were attracted to those traps, but only a few of them; there were still thousands in the honeybee colonies.

Other chemical compounds from the larvae have been identified by other research teams (Rickli et al. 1992; Donzé et al. 1998), but to date, no one has created a device for varroa control based on those attractive compounds. One possibility is to use the different compounds together, but, in fact, the challenge to use chemical compounds

Fig. 1 Experimental device to study the effect of temperature on varroa. Two of the tubes are artificially heated (*arrows*), and the mites were placed in the center of the arena. This device had been used to demonstrate the attraction by the mite to heat sources and bees

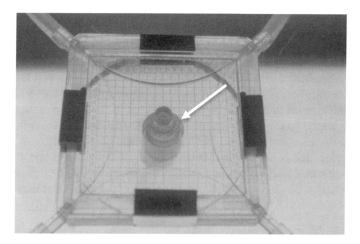

Fig. 2 Four-arm airflow olfactometer to study the behavioral response of the mite to airborne odors. *Arrow* points to where the airflow comes out

to be more attractive than the real larvae is obviously very tricky. Another possibility is to use those compounds to confuse mites (a form of control); in this case, the compounds are sprayed everywhere in the colony so that the mite does not find its target. To date, this has not been successful. Moreover, such a process could also be highly confusing to the bees as well. If these compounds are from the larva, the nurse bees must be sensitive to them. Their response to such a control might well cause nurse bees to be drawn away from the larva.

This work was a first step of a great story. Jérôme Trouiller was also working on this subject. He was in charge of the chemistry and found that the concentration of pheromone esters in brood being capped was five times higher than that produced by the youngest larvae. The hypothesis that a brood pheromone could induce the capping of the cells came naturally. To test this hypothesis, small amounts of the mixture of different compounds were applied on larvae one day before capping in the experimental brood cells, compared to control (untreated) ones. The bees capped the treated cells five times more quickly compared to the untreated controls. In addition, we tested the pheromonal effects of the esters on the capping behavior toward wax larva 'dummies.' To do this, we made larval dummies out of paraffin wax. Each of the esters was mixed in paraffin wax at different concentrations. When hot, the liquid mixture was poured in the cells of an empty comb. Once solidified, the wax dummies were pulled out and placed in empty cells in the center of a brood frame; the frame was replaced into the brood of the bees' hive (see Fig. 3a, b). After 24 h, cells containing the dummies with four of the ten esters (methyl palmitate, oleate, linoleate, or linolenate) were capped exactly as if it was a cell containing mature larvae (see Fig. 4) (Le Conte et al. 1990). It proved that those compounds, produced by the salivary glands of the larvae (Le Conte et al. 2006), represented a brood pheromone produced by the larvae as a signal to the nurse bees to cap their cells.

We ran many replicates of this capping experiment. During one of those trials, the bees capped cells containing the methyl esters, but also bees built up queen cells on the dummies containing mainly ethyl esters. The inspection of this colony revealed that it was queenless. At this time, I was collaborating with a laboratory CNRS in Marseille headed by Pr. JL Clement. This laboratory was a leader in the study of chemical communication in social insects, particularly ants and termites. They were interested in chemical signatures, blends of chemical compounds which can, for example, be produced by individuals of the colony and are recognized by nestmates. So the hypothesis that the blend of esters could be different on young and old larvae, and could be the chemical signature of the larval age, came naturally. The chemical analysis of young versus old larvae revealed a different pattern of the esters (Fig. 5). Dummies including the blend of the young larvae or of the old

Fig. 3 **a** Making the dummy larvae with a mixture of esters and paraffin. **b** Introducing dummy larvae inside brood cells of the comb

Fig. 4 In this comb, a few cells (*arrows*) containing dummy larvae with the capping pheromone have been capped exactly like cells containing real larvae

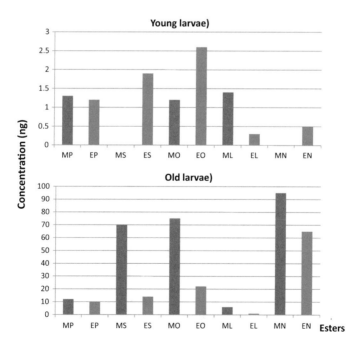

Fig. 5 Amounts of the 10 esters (ng) found on young versus old worker larvae. It is important to note that the pattern of secretion of the different compounds is different between the young and the old larvae include MP = methyl palmitate, EP = ethyl palmitate, MS = methyl stearate, ES = ethyl stearate, MO = methyl oleate, EO = methyl oleate, ML = methyl linoleate, EL = ethyl linoleate, MN = methyl linolenate, EN = ethyl linolenate

larvae were also tested. The behavioral trick used was based on the ability of a queenless colony to build up queen cells from worker cells containing young worker larvae. Queenless workers do it only on very young (less than 3 days) worker larvae, and those larvae are fed with queen jelly so that they can develop as queen and the colony is saved. We therefore introduced dummies that contained the mixture of esters corresponding to young or old larvae into empty cells of a frame removed from strong queenless colonies (Fig. 6). So if our hypothesis was correct, the bees should build queen cells on the dummies that had the young larvae mix and should cap cells with the old larvae mix. This was exactly the case (see Fig. 7), demonstrating that the modification of the blends of esters between young and old larvae represents the chemical signature of larval age to the adult worker bees (Le Conte et al. 1994).

Similar experiments, using dummies and those fatty acid esters also present in the cuticle of the queen pupae, demonstrated that a few of them are used in the recognition of queen cells by the workers (Le Conte et al. 1995a).

Fig. 6 Comb with dummy larvae containing the two different blends (corresponding to young or old larvae) to be placed into queenless bee colonies

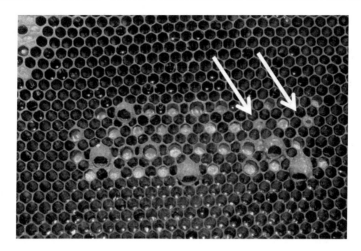

Fig. 7 Queenless bees capped the cells containing the dummy larvae with old larvae ester mix (*white arrows*) and built queen cell around the dummies with the young larvae blend of esters

Fig. 8 Wax cups used for royal jelly production and including different esters in various proportions

2   Larval Rearing

In order to find practical applications of these findings, we wanted to discover the effect those esters had on increasing royal jelly production. To produce royal jelly, we used homemade wax cells that were fixed on a stick (Fig. 8). The different molecules were mixed with the wax at different concentrations, and larvae were graphed into empty cells, following the standard methods for royal jelly production. Many replicates were done, and we found that methyl stearate increased the acceptance of the cells (more cells were reared by the bees), methyl palmitate increased the amount of royal jelly deposited in the cells, and the larvae produced in the cells with methyl linoleate were bigger, compared to controls (Le Conte et al. 1995b).

A patent has been submitted for the use of those compounds in royal jelly and queen production.

3 Effects of the Pheromone on Worker Physiology

Little was known about brood pheromones when we started our investigations. It was obvious that the brood, its age, sex, and needs (food and temperature) have to be recognized by the nurse bees so that they can provide optimal care to the immature larvae. It was also obvious that chemical communication would be a key in this communication system (Free 1987). Releaser pheromone that triggers changes in the performance behavior of the receiver should be involved; our pioneer work demonstrated it. Moreover, it was known that the brood has primer effects (modulating the physiology) on adult bees, including stimulating hypopharyngeal glands of the nurse bees (those glands provide the royal jelly which is the food for the larvae) and inhibiting development of ovaries in workers (Free 1961; Huang et al. 1989). So we hypothesized that the pheromonal blend of esters could act on those physiological processes.

First, the effect of each of the compounds, or a blend of compounds, was tested using groups of 80 bees placed in cages and exposing them to the compounds at different doses (see Fig. 9). A control set of bees were not exposed to any of the

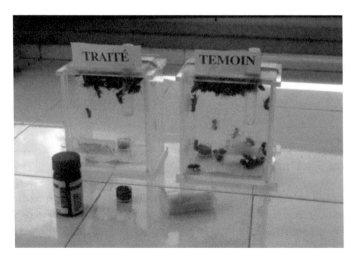

Fig. 9 Experimental cages for the testing of esters on small honeybee colonies

compounds. Ethyl oleate and methyl palmitate were found individually to stimulate the hypopharyngeal glands (Mohammedi et al. 1996), and ethyl palmitate and methyl linolenate inhibited worker ovary development (Mohammedi et al. 1998). Naturally, it may be difficult to transpose results found in cage experiments using just a few bees compared to free-flying bees. Therefore, standard triple cohort colonies were used to test the effect of those compounds. Those triple cohort colonies were made of cohorts (groups) of 500 emerging bees, 500 nurse bees, and 500 foragers. The results in caged experiments were confirmed in these experiments. The two esters stimulated hypopharyngeal glands, and two others inhibited worker ovary development (Mohammedi et al. 1998; Mohammedi 1997 thesis; Pankiw et al. 2008).

The division of labor in the honeybee colonies may also be controlled with a pheromone that inhibits development of foraging in young bees. It was known that one hormone common in insects, juvenile hormone (JH), induces foraging of the nurse bees. It has also been demonstrated that JH inhibits the development of hypopharyngeal glands (HGs) (Free 1961; Huang et al. 1989; reviewed in Robinson 1992). We hypothesized that the brood pheromone (BP) could inhibit JH level of the nurse bees so that they would keep their HGs producing royal jelly and stay home to take care of the larvae. With Pr. G. Robinson, we tested this hypothesis. The effect of BP on the JH level of caged bees was tested and was found to significantly decrease the JH level in bees exposed to BP. Again, triple cohort colonies were set up to test the effect of BP on JH level and the behavioral development of nurse bees (age at first foraging). We confirmed the effect of BP on the inhibition of JH level and found a significant effect in delaying the age at which bees leave for their first foraging (Le Conte et al. 2001). Later, we showed that BP causes the *over*expression of the honeybee gene related to nursing, while at the same time, causing the *under*expression of genes related foraging (Alaux et al. 2009).

Astonishingly, the brood pheromone (BP) has also been found to modulate the honeybee sucrose response threshold, associated with foragers choosing between pollen and nectar (Pankiw and Page 2001; Pankiw et al. 1998; Pankiw 2004, 2007). We did not find significant results on the effect of BP on pollen foraging. But using a different and clever approach, Pankiw et al. (2004, 2011) demonstrated that BP can temporarily stimulate pollen foraging by foragers. Moreover, Pankiw et al. (2004) demonstrated that BP can modulate brood-rearing behavior and behavior which influenced colony growth (Sagili and Pankiw 2009). This finding is important, as it opens the door for using BP in beekeeping to stimulate colony growth. A product, based on the ten fatty acid esters, has been tested, and the results suggest that it could be used to stimulate colony growth by stimulating protein supplement consumption during the fall (Pankiw et al. 2011; Sagili and Breece 2012).

Smedal et al. (2009) also demonstrated that BP can suppress the physiology of extreme longevity in honeybees. Feeding them BP in a supplement reduces honeybee long-term survival, the bees die quicker, probably because they had to invest in physiologic state related to feeding of the larvae. This means that, on a practical aspect, the use of those compounds in beekeeping and monitoring the colony should be done carefully.

Recent studies by Maisonnasse et al. (2009) identified a volatile compound from the young larvae, β-ocimene, which inhibited worker ovary development and stimulated worker behavioral development (Maisonnasse et al. 2010). This means first that we probably know only part of the story; second, that few chemical compounds can act together to induce a behavioral or physiological response; and third, that some compounds can have the opposite effect compared to others, to regulate social interactions in the colony.

The case of β-ocimene is interesting because it is produced by young larvae and it stimulates hypopharyngeal glands, to meet the nutritional needs of the colony (Traynor et al. 2014). In contrast, BP is produced mainly by larvae before capping, when there is no more need for food, which could be related to meet thermoregulation needs of the capped brood. The effect of β-ocimene on ovary development has been confirmed, and it has been demonstrated to stimulate the hypopharyngeal glands of the nurse bees (Traynor et al. 2015).

To conclude, it is fascinating to see how the larvae can manipulate adult bees, using chemical releaser signals that are identified by the workers (as well as their needs) and using a primer pheromone to modify the physiology of the nurse bees. Some compounds have both releaser and primer effects. The offspring of mammals produce different types of mechanical, chemical, and vibrational stimuli to manipulate their mother. These stimuli cause her to produce milk, stop ovulation, and even modify her behavior in favor of the offspring. Within the hive, the honeybee larvae are doing similar manipulations by producing those brood pheromones. While this chapter focused on the brood pheromone, many other pheromones are produced by the different actors of the colony, interacting with the bees and influencing the functions of the colony (Slessor et al. 2005; Traynor et al. 2015).

A primer pheromone, ethyl oleate (EO), has also been identified, produced by forager bees as a behavioral inhibitor of the nurse bees (Leoncini et al. 2004). The compound is produced in the esophagus from oleic acid and ethanol (from nectar of the flowers) and released by the cuticle of the bees and also, in a lesser extent, by forager regurgitates (Castillo et al. 2012a, b). Genes involved in OE formation had been identified (Castillo et al. 2012b). This finding demonstrates that the same compound, EO, is produced by different members in the colonies, the larvae and the foragers, to regulate the equilibrium between nurses and foragers, thus optimizing colony development.

The work and science produced by the study of chemical ecology are fascinating to the biologist; it can also benefit the beekeeper by understanding the mechanism of the honeybee colony and to help take better care of it. Pheromones could also be a way to optimize beekeeping productivity and represent a hope for the future. How can the beekeepers benefit from those finding? The BEEBOOST® product, made from synthetic queen mandibular pheromone, is already used successfully by beekeepers and scientists to mimic the presence of the queen. SuperBoost® has also a lot of potential to be used to stimulate colony growth, honey production, and overwintering (Sagili and Breece 2012; Lait et al. 2012). Its use by beekeepers may have practical uses in the beekeeping industry. However, the question of potential adverse effects of the use of these pheromones to disrupt the social regulations of the colony should be also studied.

References

Alaux C, Le Conte Y, Adams HA, Rodriguez-Zas S, Grozinger CM, Sinha S, Robinson GE (2009) Regulation of brain gene expression in honey bees by brood pheromone. Genes Brain Behav 8 (3):309–319. doi:10.1111/j.1601-183X.2009.00480.x

Barbier J, Lederer E (1960) Structure chimique de la substance royale de la reine d'abeille (Apis mellifera L.). C R Acad Sci, Ser 3 Sci. vie 251:1131Y1135

Butler CG, Callow RK, Johnston NC (1961) The isolation and synthesis of queen substance, 9-oxodec-trans-2-enoic acid, a honeybee pheromone. Proc R Soc Lond B Biol Sci 155:417Y432

Castillo C, Maisonnasse A, Le Conte Y, Plettner E (2012a) Seasonal variation in the titers and biosynthesis of the primer pheromone ethyl oleate in honey bees. J Insect Physiol 58:1112–1121

Castillo C, Chen H, Graves C, Maisonnasse A, Le Conte Y, Plettner E (2012b) Biosynthesis of ethyl oleate, a primer pheromone, in the honey bee (Apis mellifera L.). Insect Biochem Mol Biol 42:404–416

Donze G, SchnyderCandrian S, Bogdanov S, Diehl PA, Guerin PM, Kilchenman V, Monachon F (1998) Aliphatic alcohols and aldehydes of the honey bee cocoon induce arrestment behavior in Varroa jacobsoni (Acari: Mesostigmata), an ectoparasite of Apis mellifera. Arch Insect Biochem Physiol 37(2):129–145

Free JB (1961) Hypopharyngeal gland development and division of labour in honey-bee Apis mellifera colonies. Proc R Entomol Soc Lond A 36:1–3

Free JB (1987) Pheromones of social bees. Chapman and Hall, London

Huang Z-Y, Otis GW, Teal PEA (1989) Nature of brood signal activating the protein synthesis of hypopharyngeal gland in honey bees Apis mellifera (Apidae: Hymenoptera). Apidologie 20:455–464

Keeling CI, Plettner E, Slessor KN (2004) Hymenopteran semiochemicals. In: Chemistry of pheromones and other semiochemicals I, pp. 133–77. Springer-Verlag, Berlin

Lait CG, Borden JH, Kovacs E, Moeri OE, Campbell M, Machial CM (2012) Treatment with synthetic brood pheromone (superboost) enhances honey production and improves overwintering survival of package honey bee (Hymenoptera: Apidae) colonies. J Econ Entomol 105:304–312

Le Conte Y, Arnold G (1987) The effects of bee age and of temperature on the parasite behaviour of Varroa jacobsoni. Apidologie 18(4):305–320. doi:10.1051/apido:19870402

Le Conte Y, Arnold G (1988) A study of the thermopreferendum of Varroa jacobsoni. Apidologie 19(2):155–164

Le Conte Y, Arnold G, Trouiller J, Masson C, Chappe B, Ourisson G (1989) Attraction of the parasitic mite Varroa to the drone larvae of honey bees by simple aliphatic esters. Science 245 (4918):638–639. doi:10.1126/science.245.4918.638

Le Conte Y, Arnold G, Trouiller J, Masson C, Chappe B (1990) Identification of a brood pheromone in honeybees. Naturwissenschaften 77:334–336

Le Conte Y, Sreng L, Trouiller J (1994) The recognition of larvae by worker honeybees. Naturwissenschaften 81(10):462–465

Le Conte Y, Sreng L, Poitout SH (1995a) Brood pheromone can modulate the feeding behavior of Apis mellifera workers (Hymenoptera: Apidae). J Econ Entomol 88(4):798–804

Le Conte Y, Sreng L, Sacher N et al (1995b) Chemical recognition of queen cells by honey bee workers Apis mellifera (Hymenoptera: Apidae). Chemoecology 5/6 (1):6–12

Le Conte Y, Mohammedi A, Robinson GE (2001) Primer effects of a brood pheromone on honeybee behavioural development. Proc R Soc B-Biol Sci 268(1463):163–168

Le Conte Y, Becard JM, Costagliola G et al (2006) Larval salivary glands are a source of primer and releaser pheromone in honey bee (Apis mellifera L.). Naturwissenschaften 93 (5):237–241. doi:10.1007/s00114-006-0089-y

Leoncini I, Le Conte Y, Costagliola G, Plettner E, Toth AL et al (2004) Regulation of behavioral maturation by a primer pheromone produced by adult worker honey bees. Proc Natl Acad Sci USA 101:17559–17564

Maisonnasse A, Lenoir JC, Costagliola G, Beslay D, Choteau F, Crauser D, Becard JM, Plettner E, Le Conte Y (2009) A scientific note on E-beta-ocimene, a new volatile primer pheromone that inhibits worker ovary development in honey bees. Apidologie 40(5):562–564. doi:10.1051/apido/2009024

Maisonnasse A, Lenoir JC, Beslay D et al (2010) E-beta-ocimene, a volatile brood pheromone involved in social regulation in the honey bee colony (*Apis mellifera*). Plos One 5 (10). doi:10.1371/journal.pone.0013531

Mohammedi A, Crauser D, Paris A, Le Conte Y (1996) Effect of a brood pheromone on honeybee hypopharyngeal glands. C R Acad Sci III-Sciences De La Vie-Life Sciences 319(9):769–772

Mohammedi A (1997) Contribution a l'étude des phéromones de couvain chez l'abeille domestique *Apis mellifera* L.

Mohammedi A, Paris A, Crauser D, Le Conte Y (1998) Effect of aliphatic esters on ovary development of queenless bees (*Apis mellifera* L). Naturwissenschaften 85(9):455–458. doi:10.1007/s001140050531

Pankiw T, Sagili RR, Metz BN (2008) Brood pheromone effects on colony protein supplement consumption and growth in the honey bee (Hymenoptera: Apidae) in a subtropical winter climate. J Econ Entomol 101(6):1749–1755

Pankiw T, Page RE, Fondrk MK (1998) Brood pheromone stimulates pollen foraging in honey bees (*Apis mellifera*). Behav Ecol Sociobiol 44(3):193–198

Pankiw T, Page RE (2001) Brood pheromone modulates honeybee (*Apis mellifera* L.) sucrose response thresholds. Behav Ecol Sociobiol 49(2–3):206–213

Pankiw T (2004) Brood pheromone regulates foraging activity of honey bees (Hymenoptera: Apidae). J Econ Entomol 97(3):748–751

Pankiw T, Roman R, Sagili RR, Zhu-Salzman K (2004) Pheromone-modulated behavioral suites influence colony growth in the honey bee (*Apis mellifera*). Naturwissenschaften 91(12):575–578

Pankiw T (2007) Brood pheromone modulation of pollen forager turnaround time in the honey bee (*Apis mellifera* L.). J Insect Behav 20(2):173–180

Pankiw T, Birmingham AL, Lafontaine JP, Avelino N, Borden JH (2011) Stabilized synthetic brood pheromone delivered in a slow-release device enhances foraging and population size of honey bee, *Apis mellifera*, colonies. J Apic Res 50(4):257–264. doi:10.3896/ibra.1.50.4.02

Rickli M, Guerin PM, Diehl PA (1992) Palmitic acid released from honeybee worker larvae attracts the parasitic mite *Varroa-Jacobsoni* on a servosphere. Naturwissenschaften 79(7):320–322

Robinson GE (1992) Regulation of division of labor in insect societies. Annu Rev Entomol 37 (693):637–665

Sagili RR, Pankiw T (2009) Effects of brood pheromone modulated brood rearing behaviors on honey bee (Apis mellifera L.) colony growth. J Insect Behav 22(5):339–349

Sagili RR, Breece CR (2012) Effects of brood pheromone (superboost) on consumption of protein supplement and growth of honey bee (Hymenoptera: Apidae) colonies during fall in a northern temperate climate. J Econ Entomol 105:1134–1138

Slessor KN, Winston ML, Le Conte Y (2005) Pheromone communication in the honeybee (*Apis mellifera* L.). J Chem Ecol 31:2731–2745

Smedal B, Brynem M, Kreibich CD, Amdam GV (2009) Brood pheromone suppresses physiology of extreme longevity in honeybees (Apis mellifera). J Exp Biol 212:3795–3801

Traynor KS, Le Conte Y, Page RE (2014) Queen and young larval pheromones impact nursing and reproductive physiology of honey bee (*Apis mellifera*) workers. Behav Ecol Sociobiol 68 (12):2059–2073. doi:10.1007/s00265-014-1811-y

Traynor KS, Le Conte Y, Page RE (2015) Age matters: pheromone profiles of larvae differentially influence foraging behaviour in the honeybee, *Apis mellifera*. Anim Behav 99:1–8. doi:10.1016/j.anbehav.2014.10.009

Author Biography

Dr. Yves Le Conte, Ph.D. is Research Director at the Institut National de la Recherche Agronomique (INRA) in charge of programs dealing with behavioral, physiological, genetical aspects of the honeybee biology and pathology and the Head of the INRA Research Unit UR 406 Abeilles et Environnement, Avignon, France.

Since 1983, my research focuses on the biology and chemical ecology of honeybee colonies. With my team and collaborators, we have discovered a few pheromones from the brood and the adult bees which are at the center of social regulations in the honeybee colony. Those are primer and releaser pheromones. The primer effect had been studied at the molecular and physiological level. *Varroa destructor* is a serious threat of the honeybee in Europe, and I am also very much involved in research dealing with host parasite relationships and also applied research to control the mite.

Since the recent honeybee losses in Europe, I focus on studding the effects of different pathogens and parasites on bee health and the interactions with pesticides to understand honeybee decline from the molecular and socio-genomic level to colony level. I am also beekeeper since I was 12 years old.

Natural Selection of Honeybees Against *Varroa destructor*

Yves Le Conte and Fanny Mondet

Abstract

After varroa invaded Europe in the mid of twentieth century, a few populations of honeybee colonies have been found to survive the mite. This chapter describes the case of natural selection of honeybees in France against varroa. Different hypotheses have been tested to explain this phenomenon, such as resistance of the bees to the mite or to the associated viruses and the lower virulence of the mites. We found that the reproduction of the mite and/or the varroa sensitive hygiene are probably key factors in the survival of those bees. Other varroa resistant honeybee populations have been found in several other countries and are also described as well as the putative mechanisms of survival. Finally, we discuss the interest of those bees for scientists and beekeepers in the framework of honeybee selection and describe the successful approaches lead by scientists for honeybee selection on a specific trait.

When the varroa mite started to invade Europe in the mid of twentieth century, untreated honeybee colonies could not survive more than 1 or 3 years as the number of mites could sometimes exceed 10.000 per colony. As a result, many untreated colonies, particularly feral colonies, died. Few acaricides were used to control the mite and, as it does happen commonly in pest control, the mite became resistant to fluvalinate, a pyrethroid previously very efficient (Milani 1995). Up to now, the mite has become resistant to most of the chemical acaricides, except for amitraz in France. Having only one acaricide efficient for controlling the mite is a stressful situation which requires investments in the setting up of other acaricides with different targets to allow the rotation of treatment and so avoid varroa resistance.

Y. Le Conte (✉) · F. Mondet
INRA, UR 406 Abeilles et Environnement, Domaine Saint Paul, Site Agroparc,
CS 40 509, 84914 Avignon Cedex 9, France
e-mail: yves.le-conte@inra.fr

© Springer International Publishing AG 2017
R.H. Vreeland and D. Sammataro (eds.), *Beekeeping – From Science to Practice*,
DOI 10.1007/978-3-319-60637-8_12

Indeed, other varroa control methods have been set up, including physical or mechanical controls as well as the use of more 'natural' chemical compounds such as acids and essential oils. But those methods are usually time-consuming and with a variable efficacy. Interestingly, since the mite invaded Europe, its biology and the varroa/honeybee relationships have been extensively studied leading to the publication of many scientific articles and making this host–parasite model one of the most extensively studied (Rosenkranz et al. 2010).

Nevertheless, in 1994, feral and abandoned untreated colonies were observed to have survived for a few years in the west of France. To confirm this phenomenon, with the help of beekeepers, we collected 70 honeybee colonies which had been untreated for at least 3 years and looked well developed and healthy. Colonies collected in the north of France were placed in an apiary in the west of France close to Le Mans, and colonies collected in the south were placed in Avignon. The 70 colonies were observed for their survival, swarming, and honey production. They were not managed for honey production but just left by themselves. The queens were individually marked, and the colonies cheeked twice a month for health and development. Interestingly, those colonies survived, on average, for 6.5 years, the best being 15 years (Le Conte et al. 2007). This ability of bees to survive varroa may be due to a honeybee tolerance or resistance to the mite or its associated virus, to a lower mite virulence or to the environment.

To test the hypothesis of less virulent mites, we set up population molecular markers, mitochondrial and nuclear (microsatellites) (Solignac et al. 2003; Navajas et al. 2002) and sampled varroa mite populations in France, Europe, and few different other countries. We did not find genetic variability and concluded that the mites had a clonal population structure at this time (Solignac et al. 2005). It means that if the mites were less virulent, this would have been based on a limited number of genes. We also looked at the viruses present in the surviving bees compared to sensitive ones and found that the surviving bees had fewer viruses. We injected the bees with virus and could not find differences in survival between the two kinds of bees, suggesting that the surviving bees had fewer viruses because they had fewer varroa mites (see below), as described in Büchler et al. (2010).

Interestingly, when the mite first invaded our country, we could find high numbers of mites (up to 10.000) in the honeybee colonies and limited deformed wing virus (DWV) symptoms. A few years later we observed the opposite, i.e., lower numbers of mites in the colonies and higher DWV symptoms (Le Conte personal communication). It has been demonstrated that the varroa mite can actively modify DWV population structure in honeybee colony populations (Martin et al. 2012) as it can also do for other viruses (Mondet et al. 2014). More recently, we looked at the DWV in our resistant populations in the west of France compared to sensitive honeybee colonies and found a different recombination event between this virus and the *Varroa destructor* virus (VDV), which is very close to DWV (Dalmon et al., accepted in Scientific Reports). This virus could have evolved into a less virulent form which could explain part of the survival ability of those bees. To conclude this virus story, we must acknowledge that there are strong interactions between the mite, the viruses, and the honeybee host, and that the viruses can evolve and mutate reducing or even increasing their virulence to the bees. So nothing is fixed in those interactions which change over time.

We also tested the resistance hypothesis of those varroa surviving bees. Resistant hosts are able to maintain the parasite population at lower levels than what susceptible hosts face. Varroa population dynamics are much more important in sensitive colonies compared to our surviving colonies (Buechler et al. 2010). This may be due, at least partly, to the development of social immune strategies by the colonies. Indeed, the varroa mite reproduces less in our varroa-resistant bees compared to sensitive bees from colonies set in the same apiaries (Locke et al. 2012) and on many occasions female mites fail to effectively reproduce. This trait is known as Suppressed Mite Reproduction (SMR, see Harbo and Harris 2005). In the USA, a population has been bred for survival to varroa, initially through selection of lower mite growth in the colonies (see below). It has been shown that the colonies present the SMR trait, and this phenomenon is due to a behavior of adult bees that actively target and remove brood cells that are infested by varroa. This mechanism is called Varroa-Sensitive Hygiene (VSH) and is a specific form of hygienic behavior. However, it is possible that colonies display signs of mite reproduction failures (SMR) if infested bee larvae or pupae inhibit mite reproduction. This mechanism and VSH are being investigated in the French surviving populations and will help decipher the mechanisms underlying the SMR trait (Mondet et al. 2016).

In addition, based on the grooming behavior, we developed controlled behavioral experiments to test the ability of the bees to recognize and attack the mite. We found that resistant bees are doing much better compared to sensitive bees (Martin et al. 2001). Gene expression analysis has also revealed that the resistant bees overexpress genes related to stimuli and olfaction (Navajas et al. 2008), which fit with the fact that their antennae are more sensitive to varroa odorant compound compared to sensitive bees (Martin et al. 2001). It is interesting to notice that antennae of bees which express VSH behavior overexpress genes related to olfaction (for instance, odorant-binding proteins) (Mondet et al. 2015).

Moreover, comparing propolis harvested by sensitive or resistant colonies in the same location had shown that concentration of caffeic acid and caffeates was higher in propolis collected by our surviving colonies. Those compounds have pronounced and diverse biological properties on honeybee health (Popova et al. 2014). More studies are needed to confirm the hypothesis that surviving bees would be more capable to go to the 'pharmacy' to fight diseases.

What has happened to these bees since we published those results in 2007? Once every two years, we graft queen larvae from the three best colonies in each apiary (west and south of France) to get 20 colonies. The queens are naturally mated by local drones. About 30–35% of the colonies die within 18 months, but the rest of the colonies are good candidates for surviving to the mite, so the stock still survives efficiently.

We are focusing on the varroa mite survival, but it should be clear that those colonies are also resistant to other pathogens as they are not treated or managed against any disease. Those survival colonies swarmed (about 40% depending on the year) and similar varroa-treated colonies produced 1.7 times more honey (Le Conte et al. 2007).

The next step in this research is to investigate how those colonies would behave, regarding varroa load and honey production, when they are managed under standard beekeeping management. Preliminary trials show that they can survive in a professional beekeeping environment. Further, breeding efforts or compromises may be required to get the surviving colonies to the standard of honey production expected by the industry. Nevertheless, the surviving populations give evidence that untreated local honeybee colonies can survive the mite, which can provide an important basis for integrated varroa management. Moreover, those honeybee populations are interesting for the beekeeping industry, but also at an ecological point of view, as they are potential sources for generating feral colonies which are keys for pollination especially in areas where domestic beekeeping is not concentrated.

There are other honeybee populations naturally surviving Varroa. The best example would be the Africanized bees (AFB) which invaded the Americas from Brazil to the southern states of the USA. Initially considered as a pest because of their aggressiveness, they survive numerous stresses. It has been shown that varroa offspring mortality is a major component of this resistance phenomenon (Mondragon et al. 2006). More recently, Rivera-Marchand et al. (2012) described similar AFB populations in Puerto Rico. They are surviving the mite, but do not show similar aggressiveness. It is surprising to notice that beekeepers in America are not interested in taking advantage of those naturally varroa surviving bees as they could try to select against aggressiveness, or even import bees from Puerto Rico.

Other naturally surviving populations have been recently identified in Norway and in the Netherlands. At the moment, European scientists are exchanging queens to look at the effect of the environment on the survival of the bees (COLOSS, Ricola Foundation Program; The Persephone Charitable and Environmental Trust).

Other varroa-resistant honeybee populations have been obtained through human selection. This is the case in Sweden (Fries et al. 2006) and in France (Kefuss et al. 2004). The scientists used what they called the 'Bond' test: 'Live and let die.' Basically, they brought a large number of colonies from different strains in the same location and observed the survival. It is a slightly different approach than ours as in our case, the colonies were already observed as surviving in a local environment. This approach was successful and led to numerous scientific publications (Locke 2016).

Another step in the selection of bees resistant to the mite is to choose a trait to select for, hypothesizing that it would lead to resistant bees. As an example, this quantitative genetic approach has been successfully developed by the USDA in Baton Rouge (USA) using the SMR (Harbo and Harris 2005) and the VSH traits (Harbo and Harris 2009) as a basis for selection.

It is well known that *Apis cerana* (the original host of varroa) is varroa resistant since colonies of Asian honeybees do not die from mite infestation; we now know that it can be the case also for *Apis mellifera* as naturally Varroa surviving honeybee colonies occur in different places. Different causes can explain that phenomenon, such as individual and social immunity, olfaction, propolis, viruses, varroa reproduction, swarming. The causes may not be the same in the different bee populations, but the good news is that *Apis mellifera* can survive Varroa mite infestations without treatments. We have surviving bee populations available from naturally surviving, 'bond test' and trait-based selected populations. This is an enormous chance that we have to take and go further in selecting varroa-resistant bees from the populations we want to work with.

Different tools should be used in the future to help beekeeping on this task in the framework of IPM. One could be to identify the compounds involved in the recognition of the mite by the honeybees and use them to evaluate the ability of the colony to destroy the mites in beekeeping. We have recently identified chemicals which are good candidates in this framework. Another possibility is to search for genomic markers, as SNPs, which could be linked to the SMR and/or VSH behavior. Those could also be used as markers to select resistant bees.

References

Buechler R, Berg S, Le Conte Y (2010) Breeding for resistance to *Varroa destructor* in Europe. Apidologie 41(3):393–408

Fries I, Imdorf A, Rosenkranz P (2006) Survival of mite infested (*Varroa destructor*) honey bee (Apis mellifera) colonies in a Nordic climate. Apidologie 37(5):564–570

Harbo JR, Harris JW (2005) Suppressed mite reproduction explained by the behaviour of adult bees. J Apic Res 44(1):21–23

Harbo JR, Harris JW (2009) Responses to Varroa by honey bees with different levels of *Varroa* Sensitive Hygiene. J Apic Res 48(3):156–161

Kefuss J, Vanpoucke J, De Lahitte JD, Ritter W (2004) Varroa tolerance in France of intermissa bees from Tunisia and their naturally mated descendants: 1993-2004. Am Bee J 144(7):563–568

Le Conte Y, De Vaublanc G, Crauser D, Jeanne F, Rousselle JC, Becard JM (2007) Honey bee colonies that have survived *Varroa destructor*. Apidologie 38(6):566–572

Locke B (2016) Natural Varroa mite-surviving Apis mellifera honeybee populations. Apidologie 47(3):467–482

Locke B, Le Conte Y, Crauser D, Fries I (2012) Host adaptations reduce the reproductive success of *Varroa destructor* in two distinct European honey bee populations. Ecol Evol 2(6):1144–1150

Martin C, Provost E, Roux M, Bruchou C, Crauser D, Clement JL, LeConte Y (2001) Resistance of the honey bee, Apis mellifera to the acarian parasite *Varroa destructor*: behavioural and electroantennographic data. Physiol Entomol 26(4):362–370

Martin SJ, Highfield AC, Brettell L, Villalobos EM, Budge GE, Powell M, Nikaido S, Schroeder DC (2012) Global honey bee viral landscape altered by a parasitic mite. Science 336 (6086):1304–1306

Milani N (1995) The resistance of Varroa-jacobsoni Oud to pyrethroids—a laboratory assay. Apidologie 26(5):415–429

Mondet F, de Miranda JR, Kretzschmar A et al (2014) On the front line: quantitative virus dynamics in honeybee (Apis mellifera L.) colonies along a new expansion front of the parasite *Varroa destructor*. PLoS Pathog 10(8)

Mondet F, Alaux C, Severac D, Rohmer M, Mercer AR, Le Conte Y (2015) Antennae hold a key to *Varroa*-sensitive hygiene behaviour in honey bees. Sci Rep 5:10454

Mondet F, Kim SH, de Miranda JR, Beslay D, Le Conte Y, Mercer AR (2016) Specific cues associated with honey bee social defence against *Varroa destructor* infested brood. Sci Rep 6:25444

Mondragon L, Martin S, Vandame R (2006) Mortality of mite offspring: a major component of *Varroa destructor* resistance in a population of Africanized bees. Apidologie 37(1):67–74

Navajas M, Le Conte Y, Solignac M, Cros-Arteil S, Cornuet JM (2002) The complete sequence of the mitochondrial genome of the honeybee ectoparasite mite *Varroa destructor* (Acari: Mesostigmata). Mol Biol Evol 19(12):2313–2317

Navajas M, Migeon A, Alaux C et al (2008) Differential gene expression of the honey bee *Apis mellifera* associated with *Varroa destructor* infection. BMC Genomics 9

Popova M, Reyes M, Le Conte Y, Bankova V (2014) Propolis Chemical Composition and Honeybee Resistance against Varroa Destructor. Nat Prod Res: 1–7

Rivera-Marchand B, Oskay D, Giray T (2012) Gentle Africanized bees on an oceanic island. Evol Appl 5(7):746–756

Rosenkranz P, Aumeier P, Ziegelmann B (2010) Biology and control of *Varroa destructor*. J Invertebr Pathol 103:S96–S119

Solignac M, Vautrin D, Pizzo A, Navajas M, Le Conte Y, Cornuet JM (2003) Characterization of microsatellite markers for the apicultural pest *Varroa destructor* (Acari: Varroidae) and its relatives. Mol Ecol Notes 3(4):556–559

Solignac M, Cornuet JM, Vautrin D, Le Conte Y, Anderson D, Evans J, Cros-Arteil S, Navajas M (2005) The invasive Korea and Japan types of *Varroa destructor*, ectoparasitic mites of the western honeybee (Apis mellifera), are two partly isolated clones. Proc R Soc B-Biol Sci 272 (1561):411–419

Author Biography

Fanny Mondet has been working as a researcher at the French Institute for Agricultural Research (INRA) since 2014. She specializes in honeybee pathology. Trained as a molecular biologist and an ecologist, she completed her Ph.D. in 2014, in partnership between France and New Zealand (Pr Le Conte and Pr Mercer). Her main research focus is on the mite *Varroa destructor,* and she investigates the host–parasite interactions between honeybees and varroa. She studies the impact of the arrival of the mite on the interplay between bees, varroa, and viruses. She is also interested in developing new solutions to fight varroa, such as the ability of bees to naturally survive mite infestations. A lot of her research is dedicated to understanding *Varroa*-sensitive hygiene (VSH) behavior, at both the fundamental and applied levels.

Honeybee Venom Allergy in Beekeepers

Peter A. Ricketti and Richard F. Lockey

Abstract

Honeybees, members of the order Hymenoptera, are a major cause of systemic allergic reactions (SARs) including anaphylaxis. In certain occupations, such as beekeeping, the risk of a SAR is higher than in the general population. Beekeepers and their family members are regularly exposed to honeybee stings making them a unique population to study Hymenoptera hypersensitivity. Therefore, beekeepers and their family members need information about how to avoid stings and differentiate a local reaction from a SAR. They also need information about how and when to use an epinephrine autoinjector for a SAR and when honeybee venom immunotherapy (VIT) is indicated to prevent future SARs. For beekeepers and their family members, VIT should be given indefinitely. Once VIT maintenance is achieved, multiple monthly bee stings or optimal maintenance VIT should be continued. Alternative employment should be considered when VIT is not effective.

Disclaimer The authors have indicated there is no funding in support of this document. Additionally, there are no other relationships that might pose a conflict of interest by any of the authors. None of the authors has any professional or financial relationships relevant to the subject matter in this paper. This manuscript and any tables or figures therein have not been submitted to or are under consideration by any other publisher or publication.
Support There are no sources of support that require acknowledgment.
This manuscript was prepared by Microsoft Word on a PC platform.

P.A. Ricketti (✉) · R.F. Lockey
Division of Allergy and Immunology, Department of Internal Medicine,
University of South Florida Morsani College of Medicine and
James A. Haley Veterans Hospital, Tampa, FL, USA
e-mail: rlockey@health.usf.edu

© Springer International Publishing AG 2017
R.H. Vreeland and D. Sammataro (eds.), *Beekeeping – From Science to Practice*,
DOI 10.1007/978-3-319-60637-8_13

Keywords

Hymenoptera · Apidae · Honeybee · Beekeeper · Anaphylaxis · Systemic allergic reaction · Large · Local reaction · Skin test · Intradermal test · Venom-specific IgE · Systemic mastocytosis · Sting challenge · Epinephrine · Venom immunotherapy

1 Introduction

The order Hymenoptera is one of the largest groups of insects and includes wasps, hornets, yellow jackets, ants, and bees. Hymenopteran stings are a major cause of systemic allergic reactions (SARs) and anaphylaxis (Rueff et al. 2011). Beekeepers and their family members are especially susceptible given their increased exposure to honeybees, making them a unique population to study Hymenoptera hypersensitivity (Bilo et al. 2012). Furthermore, because of the high degree of honeybee sting exposure, the indications and protocols for venom immunotherapy (VIT) may differ from other individuals who are not regularly exposed to these stings (Muller 2005). This review focuses on the epidemiology and occupational aspects of honeybee venom allergy, its pathogenesis, clinical

Table 1 Clinical criteria for diagnosing anaphylaxis

Anaphylaxis is highly likely when any *one* of the following 3 criteria is fulfilled:
1. Acute onset of an illness (minutes to several hours) with involvement of the skin, mucosal tissue, or both (e.g., generalized hives, pruritus or flushing, swollen lips–tongue–uvula)
And At Least One of the Following
a. Respiratory compromise (e.g., dyspnea, wheeze–bronchospasm, stridor, reduced PEF, hypoxemia)
b. Reduced BP or associated symptoms of end-organ dysfunction (e.g., hypotonia [collapse], syncope, incontinence)
2. Two or more of the following that occur rapidly after exposure *to a likely allergen for that patient* (minutes to several hours):
a. Involvement of the skin–mucosal tissue (e.g., generalized hives, itch–flush, swollen lips–tongue–uvula)
b. Respiratory compromise (e.g., dyspnea, wheeze–bronchospasm, stridor, reduced PEF, hypoxemia)
c. Reduced BP or associated symptoms (e.g., hypotonia [collapse], syncope, incontinence)
d. Persistent gastrointestinal symptoms (e.g., crampy abdominal pain, vomiting)
3. Reduced BP after exposure to *known allergen for that patient* (minutes to several hours):
a. Infants and children: low systolic BP (age specific) or greater than 30% decrease in systolic BP*
b. Adults: systolic BP of less than 90 mm Hg or greater than 30% decrease from that person's baseline

Table 2 Proposed modification of the 2010 WAO grading system

Grading system for SARs

Grade 1	Grade 2	Grade 3	Grade 4	Grade 5
			Anaphylaxis	
Symptoms(s)/sign(s) from 1 organ present	Symptoms(s)/Sign(s) from ≥ 2 organ symptoms listed in grade 1	**Lower airway**	**Lower airway**	**Lower or upper airway**
Cutaneous		- Mild bronchospasm, e.g., cough, wheezing, shortness of breath which corresponds to treatment	- Severe bronchospasm, e.g., not responding or worsening in spite of treatment	- Respiratory failure and/or
- Urticaria and/or erythema warmth and/or pruritis, other than localized at the injection site		And/or	And/or	**Cardiovascular**
And/or		**Gastrointestinal**	**Upper airway**	-Collapse/hypotension[†]
- Tingling, or itching of the lips* or		- Abdominal cramps* and/or vomiting/diarrhea	- Laryngeal edema with stridor	And/or
- Angioedema (not laryngeal)		**Other**	- Any symptoms(s)/sign(s) from grades 1 or 3 would be included	- Loss of consciousness (vasovagal excluded)
Or		- Uterine cramps		- Any symptom(s)/sign(s) from grades 1, 3, or 4 would be included
Upper respiratory		- Any symptom(s)/sign(s) from grade 1 would be included		
- Nasal symptoms (e.g., sneezing, rhinorrhea, nasal pruritis, and/or nasal congestion)				
And/or				
- Throat-clearing (itchy throat)*				
And/or				
- Cough not related to bronchospasm				
Or				
Conjunctival				
- Erythema, pruritis, or tearing				
Or				

(continued)

Table 2 (continued)

Grading system for SARs

Grade 1	Grade 2	Grade 3	Grade 4	Grade 5
			Anaphylaxis	

Other
Nausea
Metallic Taste

The final grade of the reaction is not determined until the event is over, regardless of the medication administered to treat the reaction. The final report should include the first symptom(s)/sign(s) and the time of onset after the causative agent exposure and a suffix reflecting if and when epinephrine was or was not administered: a, ≤ 5 min; b, >5 to ≤ 10 min; c, >10 to ≤ 20 min; d, >20 min; z, epinephrine not administered

Final report: Grade 1–5; a–d, or z; first symptom(s)/sign(s); time of onset of first symptom(s)/signs(s). Case example. Within 10 min of receiving an AIT injection, a patient develops generalized urticaria followed by a tickling sensation in the posterior pharynx. Intramuscular epinephrine is administered within 5 min of symptoms(s)/sign(s) resulting in complete resolution of the reaction. The final report would be Grade 2; a; urticaria; 10 min

*Application-site reactions would be considered local reactions. Oral mucosa symptoms, such as pruritus, after SLIT administration, or warmth and/or pruritus at a subcutaneous immunotherapy injection site would be considered a local reaction. However, tingling or itching of the lips or mouth could be interpreted as a SAR if the known allergen, e.g., peanut, is inadvertently placed into the mouth or ingested in a subject with a history of a peanut-induced SAR. Gastrointestinal tract reactions after SLIT or oral immunotherapy (OIT) would also be considered local reactions, unless they occur with other systemic manifestations. SLIT or OIT reactions associated with gastrointestinal tract and other systemic manifestations would be classified as SARs. SLIT local reactions would be classified according to the WAO grading system for SLIT local reactions.[6] A fatal reaction would not be classified in this grading system but rather reported as a serious adverse event

†Hypotension is defined per the National Institute of Allergy and Infectious Disease/Food Allergy and Anaphylaxis Network Expert Panel criteria[3]. "Reduced blood pressure after exposure to known allergen for that subject (minutes to several hours)"

A) Infants and children: low systolic blood pressure (age-specific) or greater than 30% decrease in systolic blood pressure

Low systolic blood pressure for children is defined as follows

• 1 mo to 1 y: < 70 mm Hg
• 1–10 y: < 70 mm Hg + [2 × age]
• 11–17 y: < 90 mm Hg

B) Adults: systolic blood pressure of less than 90 mm Hg or greater than 30% decrease from that person's baseline

Table 3 Taxonomy of the Hymenoptera insect order

Family and subfamily	Scientific name	Common name
Apidae	*Apis mellifera*	Honeybee
	Bombus spp.	Bumblebee
	Megabombus spp.	
	Halictus spp.	Sweatbee
	Dialictus spp.	
Vespidae		
Vespinae	*Vespula* spp.	Yellow jacket
	Dolichovespula arenaria	Yellow hornet
	Dolichovespula maculata	White-faced hornet
Polistinae	*Polistes* spp.	Paper wasp
Formicidae	*Solenopsis invicta*	Fire ant
	Myrmecia spp.	Jack jumper ant
	Pogonomyrmex spp.	Harvester ant
	Pachycondyla spp.	

Golden (2009)

features, diagnosis, and means by which to prevent sting-induced SARs. It also covers treatment for beekeepers who have SARs to honeybee stings and the indications for the use of VIT.

2 Definitions of a Systemic Allergic Reaction (SAR) and Anaphylaxis

Anaphylaxis is a severe, potentially fatal SAR that may occur suddenly after contact with an allergen (Sampson et al. 2006). Sampson et al. (2006) established criteria for an anaphylactic reaction as outlined in Table 1, and the World Allergy Organization devised a grading system for SARs, with anaphylaxis being included in grades 3 and 4 (Cox et al. 2010). A modification of this grading system was published in 2017 (Table 2; Cox et al. 2017). Given the lack of specific criteria for diagnosing a SAR versus anaphylaxis, the term SAR will be used throughout this manuscript.

3 Taxonomy

Honeybees belong to the order Hymenoptera, and only the female honeybees can sting as their stinging apparatus is a modified ovipositor not capable of egg laying (Casale and Burks 2014). Although some stinging insect species sting offensively to disable and capture prey, most stings from honeybees occur when defending their

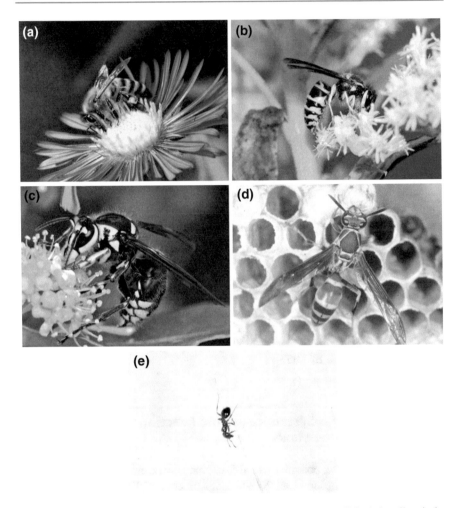

Fig. 1 Stinging insects of the order Hymenoptera. (**a** honeybee (*Apis mellifera*), **b** yellow jacket (*Vespula maculifroms*), **c** white-faced hornet (*Dolichovespula maculata*), **d** paper wasp (*Polistes exclamans*), **e** imported fire ant (*Solenopsis invicta*). Golden 2009)

nests. Three different taxonomic families of Hymenoptera insects exist as follows: Apidae (honeybees, bumblebees, and sweat bees), Vespidae (hornets, wasps, and yellow jackets), and Formicidae (ants) (Table 3; Fig. 1; Golden 2009).

Feral honeybee nests can be found in tree hollows, in old logs, or in buildings. Hives usually contain thousands of honeybees; however, during the past ten years, colony numbers have declined attributed to a host of new pathogens and pesticides (Golden 2009; Golden et al. 2011). Domesticated honeybees in commercial hives are used for their honey production. They are bred to be relatively docile and non-aggressive. The sting apparatus of a honeybee is barbed and, when inserted into the skin, pulls away from the body when the bee flies off the skin, eviscerating it, and causing its death (Golden 2009). The presence of a stinger is usually indicative of a honeybee sting, but not always, because the sting apparatus of ground-nesting yellow jackets also is barbed and can remain in the skin, although the latter is rare (Golden et al. 2011). Other than beekeepers, most honeybee stings occur in children and others who go barefoot outdoors or who handle flowering plants without gloves. Even though honeybees are more docile than other species of the order Hymenoptera, their stings are more likely to sensitize and cause a subsequent SAR (Casale and Burks 2014).

Africanized honeybees, or so-called killer bees, are hybrids that resulted from the interbreeding in South America of domestic and African honeybees (Golden et al. 2011). They entered the USA through the southern border during the 1990s and now are present in Texas, New Mexico, Arizona, Nevada, and California (Golden 2009). The venom from Africanized honeybees is identical to the European honeybee venom. However, these honeybees are more aggressive and increase the incidence of life-threatening SARs or fatal toxic reactions (Golden 2009; Golden et al. 2011). In rare instances, delayed reactions to stings occur, the mechanism of which is unknown. Serum sickness, encephalitis, peripheral and cranial neuropathies, glomerulonephritis, myocarditis, and Guillain–Barre syndrome may also be a consequence of stings from honeybees (Golden 2009; Golden et al. 2011).

Bumblebees are usually less aggressive than honeybees and rarely sting. Stings from bumblebees usually occur in high-risk areas such as greenhouses where they are used to pollinate commercially grown flowers. There is little or no cross-reactivity between honeybee and bumblebee venoms (Golden 2009). Sweat bees are named for their attraction to salt in human perspiration. Their colonies predominantly exist in North America, and these bees are commonly found in flowering gardens or meadows. Additionally, sweat bee sting-induced SARs are rare, and there is little or no cross-reactivity between their venom and honeybee venom.

4 Epidemiology

SARs to Hymenoptera stings occur in 1–4% of the population at large in all age-groups as indicated by various European and US publications (Charpin et al. 1989; Golden et al. 1989; Settipane and Boyd 1970; Struppler et al. 1997). They cause at least 40 reported deaths per year in the USA; however, estimates are greater because of unreported deaths (Freeman 2004; Golden 2009). Up to 31% of beekeepers and their families report large local reactions (LLRs) and 14–32% SARs following honeybee stings (Muller 2005). Additionally, 30–60% of beekeepers have venom-specific serum IgE. Annila et al. (1997) and Muller et al. (1977), Muller (2005) demonstrated that a positive skin test to honeybee venom is not diagnostic of a history of a SAR in this cohort and that sting reactions are highest during the first years of beekeeping. The risk of honeybee sting allergy increases with exposure; therefore, beekeepers have an increased risk of a sting-induced SAR, the longer they practice their occupation (Eich-Wanger and Muller 1998).

Bousquet et al. (1984) demonstrated that infrequently stung beekeepers are at highest risk to develop a SAR. His studies show that SARs occurred in 45% of beekeepers who experienced fewer than 25 annual stings, whereas none occurred with greater than 200 annual stings (Bousquet et al. 1984; Muller 2005). These reactions occur most frequently during the spring but also during the fall when honey is first harvested, and beekeepers resume working with their colonies. Additionally, SARs are common in beekeepers who have been at their profession the longest (Muller 2005). The risk of a SAR may be greater in atopic vs non-atopic beekeepers, the former of which have a heightened immune response to common allergens and a predisposition to develop allergic diseases such as allergic rhinitis, asthma, and atopic dermatitis. Some studies also indicate that atopy predisposes a person to an increased risk of a honeybee sting-induced SAR. For example, a 2006 Turkish study demonstrated an 11-fold increase in the risk of a SAR in beekeepers with two or more concurrent atopic diseases (Celikel et al. 2006). Another example is that beekeepers who have allergic rhinoconjunctivitis and/or asthma are more easily sensitized than non-atopic beekeepers, either through the inhalation of honeybee dust debris or because of multiple stings (Miyachi et al. 1979).

Predisposing factors for a SAR in a 2011 British beekeeper study by Richter et al. (2011) include females, a family member with honeybee venom allergy and greater than 2 years of beekeeping before a SAR occurs. In this study, only 44% of beekeepers with a SAR went to an emergency department (ED). Of these, 66% were evaluated by an allergist/immunologist, but only 18% carried an epinephrine auto-injector (Richter et al. 2011), suggesting a need to further educate this high-risk group about being appropriately evaluated for VIT and use of epinephrine.

5 Hymenoptera Venom Composition and Honeybee Venom

Knowledge of venom composition and structures of venom allergens is important to understand how to accurately diagnose and treat insect venom allergy (Bilo et al. 2012). Dr. Mary Loveless challenged the conventional belief that successful immunotherapy of Hymenoptera-sensitive individuals could be obtained using whole body extracts derived from the whole honeybee versus just the venom in the early 1950s. She championed the idea that the allergens were concentrated in the venom and hypothesized that venom therapy would prove more efficacious than whole body extracts (Loveless and Fackler 1956). Twelve insect-sensitive persons were treated with multiple intradermal (ID) injections of venom sac extracts over the course of one or two days once annually, receiving a targeted cumulative dose of six sacs of venom. Immunity was assessed by deliberate sting challenges. It was uniformly successful (Levine and Lockey 2003; Loveless and Fackler 1956). Dr. Loveless, in a subsequent study, reported successful use of venom emulsion injections and even periodic

Table 4 Honeybee venom allergens

Allergen	Biochemical name	Molecular weight (kDa)	Major/minor
Api m 1	Phospholipase A2	16	Major
Api m 2	Hyaluronidase	39	Major (?)
Api m 3	Acid phosphatase	43	Minor
Api m 4	Melittin	2.8	Minor
Api m 5	Low dipeptidyl peptidase IV	102	?
Api m 6	Cysteine-rich trypsin inhibitor	8	Minor
Api m 7	CUB serine protease	39	?
Api m 8	Carboxylesterase	70	Minor
Api m 9	Serine carboxylesterase	60	?
Api m 10	Icarapin	50–55	?
Api m 11	Major royal jelly protein	50	?
Api m 12	Vitellogenin	200	?

'?' indicates that it is not yet known whether the allergen is major or minor
Allergens of honeybee venom characterized by molecular weight and major vs minor determinants. The most important honeybee venom allergen is Phospholipase A2, also referred to as *Api m 1*. Major/minor refers to whether each honeybee venom allergen is a major or minor allergen
Bilo et al. (2012)

deliberate stings alone to control insect sting allergy (Levine and Lockey 2003; Loveless 1962). However, her work was not scientifically based in that it was not double-blind controlled. Whole body extracts were not replaced with venom extracts until a double-blind controlled study by Hunt et al. (1976) proved that VIT was more effective than either yellow jacket or honeybee whole body extracts or placebo.

Honeybee venom contains 12 known allergens (Table 4; Bilo et al. 2012). The most important allergen is phospholipase A2 (PLA2), also referred to as *Api m* 1, which is a 134 amino acid glycoprotein (Bilo et al. 2012). PLA2, in addition to eliciting an IgE response, may increase production of biologically active compounds known as leukotrienes from the breakdown of white blood cells. These leukotrienes may act as adjuvants and enhance a beekeeper's susceptibility to develop a SAR.

6 Pathogenesis of Honeybee Venom Allergy and Venom Immunotherapy

An allergy is a medical condition that can occur any time in a person's life. This involves an abnormal reaction to an ordinarily harmless substance, in this case, an allergen. The host's immune system views the allergen as non-self or an invader, and a chain reaction is initiated with increased production of specific IgE antibodies. These specific IgE antibodies attach to mast cells along venules and

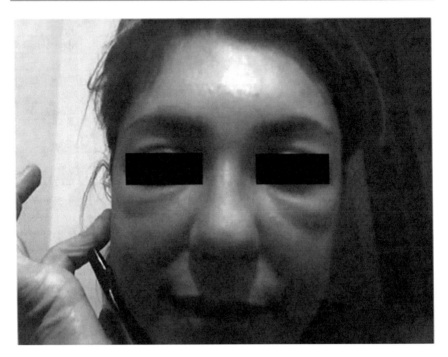

Fig. 2 Facial swelling from a large local reaction (LLR) (Severino et al. 2009)

arterioles and mucosal membranes throughout the body. When two IgE molecules are bridged by an allergen, in this case, a venom allergen, they release potent chemical mediators such as histamine. If the subject is severely sensitized, a SAR may occur, prompting a possible requirement for emergency medical care. Another example of a common allergen is ragweed pollen, which, in this case, is inhaled into the nose. If the subject is sensitized or allergic to ragweed, two IgE molecules located on the mast cell are bridged by the ragweed allergen inducing a "naso-ocular" allergic reaction manifested by runny nose, sneezing, nasal itching, nasal stuffiness and itchy, red, tearing of the eyes.

Thus, SARs to honeybee stings are secondary to a type I IgE-mediated hypersensitivity reaction. Immediate type skin tests, by either the prick puncture or ID routes, with honeybee venom are positive in more than 95% of subjects who have had a SAR. Likewise, specific IgE, as measured in the serum, is detected in more than 90% of subjects during the first year, following a sting-induced SAR (Celikel et al. 2006).

High levels of honeybee venom-specific IgG and IgG_4 antibodies occur in heavily exposed beekeepers and in individuals who are on honeybee VIT (Bilo et al. 2012). High levels of venom-specific IgG_4 occur in heavily exposed beekeepers who are repeatedly restung and during prolonged VIT, suggesting that this antibody competitively inhibits IgE binding to an allergen, in this case, honeybee venom. This mechanism explains how these two groups, i.e., those who are frequently stung and those on VIT, develop natural immunity or tolerance to a venom

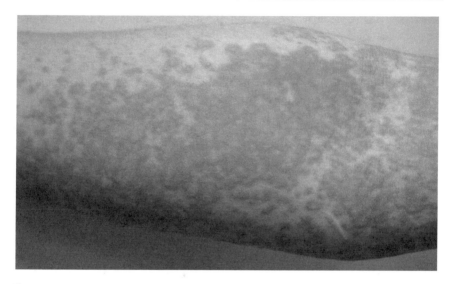

Fig. 3 Generalized urticaria of the upper extremity (Heilman 2010)

sting and thus are protected from future SARs compared to infrequently stung beekeepers and those not on VIT (Pesek and Lockey 2014).

7 Clinical Features

Most individuals stung by a honeybee experience localized swelling, some of which are small, the size of a dime or nickel, or large, ≥ 10 cm. A minority develop excessive, localized swelling, which can be "immediate" and peak within 1–2 h or "delayed" and resolve over several days (Golden 2015). LLRs are commonly mistaken for cellulitis as the intense inflammation associated with this reaction may cause apparent lymphangitis, inflammation of the lymphatic vessels, within the first 24–48 h. These reactions are usually not serious although a LLR with severe swelling occurring in the mouth or upper airway can be life-threatening (Fig. 2; Golden 2009; Golden et al. 2011; Severino et al. 2009).

A small percentage of individuals stung by honeybees develop SARs. For example, a beekeeper is stung on the foot and has signs and symptoms of a SAR including generalized urticaria, erythema, itching, and other symptoms (Fig. 3; Heilman 2010). Acute SARs typically occur rapidly following a Hymenoptera sting but rarely are delayed for several hours (Casale and Burks 2014; Golden 2015). Rarely, biphasic reactions may occur. A biphasic reaction is an immediate reaction that occurs within minutes followed by a recurrence of the same symptoms and signs of a SAR 6–8 h later (Golden 2009; Golden et al. 2011). Factors associated with an increased risk of a SAR include a honeybee sting, with a greater risk than other Hymenoptera stings; underlying mast cell disorders; a previous SAR;

preexisting cardiovascular disease; and possibly concomitant treatment with a beta-blocker. Beta-blockers may block the beta agonist effects of epinephrine, the drug of choice to treat a SAR, thus reducing the therapeutic effect of epinephrine. Concomitant treatment with angiotensin-converting enzyme (ACE) inhibitors is controversial (Freeman 2004; Golden 2009; Golden et al. 2011). ACE inhibitors prevent the breakdown of neuropeptides and bradykinin released as a by-product of mast cell degranulation (Casale and Burks 2014). This could explain the fact that there is some evidence, which is inconclusive, of an increased number of SARs in subjects on ACE inhibitors. However, such a risk was not demonstrated by Sto-evesandt et al. (2014) in their study of 743 Hymenoptera-allergic individuals. When in doubt, an angiotensin receptor blocker can be substituted for an ACE inhibitor. Beekeepers with a history of hypertensive cardiovascular disease should discuss the risks of these medications with their cardiologist in the event they develop a SAR from a sting.

SARs present with a spectrum of symptoms and signs and can affect multiple organ systems including the skin, gastrointestinal tract, nervous system, and both the upper and lower respiratory tracts. Hallmarks of a SAR include the involvement of one or more organ systems and the development of hypotension or respiratory distress (Table 1; Sampson et al. 2006). In general, when symptoms of a SAR develop rapidly, the end result has the potential to be more severe. Death from such reactions typically results from upper airway obstruction, respiratory failure, and hypotension. Beekeepers must be aware of these symptoms and signs and appropriately treat themselves with an epinephrine auto-injector.

8   Referral and Diagnosis

An accurate diagnosis of a honeybee sting-induced SAR in beekeepers is essential given its potential consequences. This includes providing epinephrine auto-injectors for beekeepers, VIT, and in rare cases, changing an occupation or living location (Golden 2015). An accurate history of the sting reaction is most important. The clinician should document each reaction and its severity and, if possible, identify the culprit insect. Helpful observations include the time, location, and circumstances of the sting; the presence or absence of a stinger in the skin; and previous sting history and possible specimens collected (Golden 2015). Knowledge of regional variation in species also may be helpful since most stings occur during the warmer months. If possible, case records should be reviewed for objective manifestations of a SAR. When a SAR is suspected, beekeepers should be referred to an allergist/immunologist for appropriate skin and/or in vitro blood testing and possible treatment. VIT, when indicated, will depend on the history of a SAR and the results of appropriate skin or *in vitro* tests. Education, prevention, and appropriate use of self-administered epinephrine auto-injectors are essential for beekeepers.

9   Skin and In Vitro Tests for Venom-Specific IgE

Skin prick and ID skin tests or *in vitro* venom-specific IgE tests are necessary to confirm sensitivities for beekeepers with a suspected history of a SAR secondary to a honeybee sting (Golden 2009; Golden et al. 2011). The degree of specific IgE sensitivity does not correlate with the severity of the sting reaction. Some subjects who have had near fatal SARs may only react to the highest concentration of venom used for this purpose. Others, who have had LLRs, with no essential risk for a SAR, also may have positive skin tests. Lockey et al. (1989) performed a 3-year Hymenoptera venom study from 1979 to 1982 based on case histories and venom skin test results of 3236 subjects who had reacted adversely with a SAR or LLR to stings of Hymenoptera insects. This study indicates that there are no significant differences of the wheal and erythema sizes associated with different venoms or different historical sting reactions. Therefore, the history of the SAR as well as skin test or in vitro venom-specific IgE results is important to document before recommending VIT.

For beekeepers who have a convincing history of a SAR, but with negative skin tests, *in vitro* venom-specific IgE tests are necessary (Golden 2009; Golden et al. 2011). Similarly, negative *in vitro* venom-specific IgE tests should be followed up with venom skin tests. Venom skin tests and venom-specific IgE assays correlate imperfectly, and neither test is superior to confirm the presence of Hymenoptera venom-specific IgE.

False-negative skin or serum-specific IgE tests may occur within the first few weeks after an insect sting reaction. This is attributed to a "refractory period" theoretically explained by the fact that specific IgE to venoms has been exhausted during the reaction and has not yet been replaced by the immune system. Therefore, tests preferably should be carried out six weeks after a sting-induced reaction (Casale and Burks 2014; Golden 2009).

10   Management of LLRs and SARs

The vast majority of LLRs do not require treatment other than symptomatic therapy with cold compresses, an analgesic, an oral H_1-anthistamine, and a topical glucocorticoid ointment to reduce itching and localized pain and swelling (Freeman 2004; Golden 2009; Golden et al. 2011). If a LLR occurs on the face or compromises the airway of a beekeeper, then epinephrine, an antihistamine, and a glucocorticosteroid are indicated. VIT reduces the size and duration of LLRs in beekeepers given their unavoidable exposure.

LLRs do not predispose to SARs, and the risk of a SAR in subjects with a history of a LLR is less than 5–10% (Golden et al. 2011). Therefore, intramuscular (IM) epinephrine is not recommended for LLRs. However, epinephrine for IM use and instructions on its administration is indicated for subjects with a history of a SAR. A minimum of two epinephrine auto-injectors for IM use also should be prescribed for high-risk subjects. These include Hymenoptera-sensitive beekeepers, those with systemic mastocytosis and those who have an elevated baseline serum tryptase level without systemic mastocystosis (Golden et al. 2011; Haeberli et al. 2003; Rueff et al. 2011).

Mastocytosis is a rare disease characterized by increased mast cells in the skin and/or internal organs, and the incidence of this disease ranges between 1 and 7.9% in venom-allergic subjects (Rueff et al. 2006). Subjects with Hymenoptera venom allergy who suffer from mastocytosis develop SARs more frequently than those who do not, and physicians should consider mastocytosis in anyone with a SAR to a Hymenoptera sting, especially with negative venom testing (Bonadonna et al. 2014). Several deaths from sting reactions were reported in subjects with masto-cytosis after VIT was stopped. Therefore, beekeepers with mastocytosis and ele-vated serum tryptase levels without mastocytosis should remain on VIT indefinitely (Rueff et al. 2006).

Treatment of Hymenoptera-induced SARs is identical to the treatment of a SAR caused by any other allergen. The beekeeper needs to administer IM epinephrine into the mid-anterolateral thigh and be transported to the nearest ED at the first symptom or sign of a SAR. The recommended dose of epinephrine 1:1000 (1 mg/mL) solution is 0.01 mg/kg of body weight, with a maximum single dose of 0.5 mg in an adult and 0.3 mg in a child (Casale and Burks 2014; Golden 2009). FDA-approved epinephrine auto-injectors contain 0.15 or 0.30 mg of epinephrine. The physician should prescribe the most appropriate auto-injector based on the weight of the affected subject. For beekeepers who experience persistent or refractory symptoms, it should be repeated as often as necessary or at intervals of 5-15 min. Immediate delivery of epinephrine is paramount since delay in its administration may contribute to more serious reactions and even a fatality. Therefore, there is no absolute contraindication for the use of epinephrine for a SAR.

Education for beekeepers is of paramount importance particularly for the use of self-injectable epinephrine. Beekeepers with a history of a SAR should carry sev-eral epinephrine auto-injectors at all times, and if they are stung and know they are sensitive, use it immediately or with the earliest onset of SAR symptoms or signs. Several epinephrine auto-injectors are ideal in case the SAR requires more than one injection. The number of epinephrine auto-injectors prescribed should be discussed between beekeepers and their physicians.

11 Risk Factors and Prevention

Potential risk factors for SARs in beekeepers include the following: 1. The first years of beekeeping and the first stings in the spring; 2. fewer than 25 annual bee stings; 3. high skin test sensitivity and specific IgE to honeybee venom and low serum venom-specific IgG_4; 4. history of atopic disease; and 5. symptoms of upper respiratory tract allergy while working at the honeybee hive (Anilla et al. 1997;

Bousquet et al. 1984; Eich-Wanger and Muller 1998; Miyachi et al. 1979; Muller et al. 1977). Highly exposed beekeepers should wear protective clothing with nets, especially during hazardous activities. Thoughtful breeding of less aggressive honeybees also should be considered. Mauss (2008) described how less aggressive honeybees existed for prolonged periods in the German locale as "sweet temper" was used as a main selection criterion to breed honeybees. This reduced their aggressive behavior as well as the number of stings and SARs. "Sweet-tempered" honeybees produce less honey and may not be as good at pollination, a trade-off from their less aggressive behavior.

12 Venom Immunotherapy

Honeybee VIT is indicated in beekeepers with a history of a SAR and a desire to continue working as a beekeeper (Golden et al. 2011). It not only reduces morbidity and mortality but also improves the quality of life, thus reducing the anxiety of venom-allergic individuals, making such therapy perhaps more cost-effective.

The efficacy of VIT has been confirmed by sting challenges and field stings in prospective controlled and uncontrolled studies (Bilo et al. 2012). Initial VIT injections are usually administered once or twice weekly, beginning with doses no greater than 0.1–1.0 μg and gradually increasing the dose to a maintenance of 100 μg of each insect venom to which a beekeeper is sensitized. Maintenance dose injections can be increased to 4-week intervals during the first 12–18 months and, thereafter, to a 6-week interval for an additional 12–18 months and then subsequently administered every 8 weeks (Bonifazi et al. 2005). The maintenance dose of 100 μg was selected in early clinical trials because it was estimated to be equivalent to two honeybee stings (50 μg per sting) (Golden et al. 2011). However, since VIT with honeybee venom is less effective than VIT with vespid venom, new recommendations include treating subjects with honeybee venom allergy who are at an increased risk because of high exposure, such as beekeepers, with a maintenance dose of honeybee venom, 200 μg (Goldberg and Confino-Cohen 2010; Muller et al. 1992). Safe and more accelerated VIT schedules are available for beekeepers who wish to achieve rapid maintenance doses because of the increased risk of SARs due to their occupation. VIT should only be prescribed by trained physicians who have experience and knowledge of its administration and the treatment of SARs, the latter of which can occur as a side effect (Golden 2009; Golden et al. 2011).

Although guidelines indicate that VIT may be discontinued after 3–5 years, there is a greater chance of SAR relapse in beekeepers (Golden 2009; Golden et al. 2011). Despite this, many beekeepers refuse to stop their work because it is their occupational hobby and, many times, a source of income. Beekeepers should therefore continue VIT indefinitely, or if necessary, if they are still having SARs, with a higher monthly maintenance for the duration of their professional activity (Bilo et al. 2012; Muller 2005; Rueff et al. 2011). If beekeepers want to stop VIT, sting provocation tests under controlled conditions should be performed and the sting tolerated before they resume beekeeping (Bilo et al. 2012; Fischer et al. 2013; Pesek and Lockey 2014). Although sting challenges are not used for diagnostic purposes and because there are concerns regarding the reproducibility of a sting, those who successfully pass a challenge have improvements in their quality of life and are more confident about discontinuing VIT (Freeman 2004). Studies show that once maintenance injections and field stings are tolerated for 3 years, beekeepers may receive one or two weekly stings or up to four monthly stings at the hive as a way of replacing VIT (Bilo et al. 2012; Eich-Wanger and Muller 1998; Muller 2005). Additionally, monthly maintenance injections of 200 μg of VIT may be resumed during the winter months when the risk of a SAR is highest (Bilo et al. 2012; Eich-Wanger and Muller 1998; Muller 2005). If beekeepers cannot tolerate weekly or monthly sting challenges or maintenance VIT and continue to experience SARs, then these individuals should strongly consider giving up their profession.

13 Conclusion

Hymenoptera venom allergy causes SARs, and in certain occupations, such as beekeeping, the risk of a major SAR is higher than the average population. Therefore, beekeepers who have a history of a SAR should be instructed to keep epinephrine auto-injectors with them at all times and use it immediately when they are stung or with the earliest onset of a SAR. Allergic beekeepers also should be treated with a VIT maintenance dose of 200 μg, which gives increased protection from subsequent stings. VIT should be continued indefinitely for sensitive bee-keepers as long as honeybee exposure remains high. If VIT or sting challenges are unsuccessful or cannot be performed, the beekeeper should consider an alternative employment.

Abbreviations

ACE	Angiotensin-converting enzyme
ED	Emergency department
ID	Intradermal
IM	Intramuscular
LLR	Large, local reaction
PLA2	Phospholipase A2
SAR	Systemic allergic reaction
VIT	Venom immunotherapy

References

Annila IT, Annila PA, Morsky P (1997) Risk assessment in determining systemic reactivity to honeybee stings in beekeepers. Ann Allergy Asthma Immunol 78:473–477

Bilo BM, Antonicelli L, Bonifazi F (2012) Honeybee venom immunotherapy: certainties and pitfalls. Immunotherapy 4(11):1153–1166

Bonadonna P, Lombardo C, Zanotti R (2014) Mastocytosis and allergic diseases. J Investig Allergol Clin Immunol 24:288–297

Bonifazi F, Jutel M, Bilo BM et al (2005) Interest group on insect venom hypersensitivity: prevention and treatment of hymenoptera venom allergy: guidelines for clinical practice. Allergy 60:1459–1470

Bousquet J, Menardo JL, Aznar R et al (1984) Clinical and immunologic survey in beekeepers in relation to their sensitization. J Allergy Clin Immunol 73:332–340

Casale TB, Burks AW (2014) Hymenoptera-sting hypersensitivity. N Engl J Med 370:1432–1439

Celikel S, Karakaya G, Yurtsever N et al (2006) Bee and bee products allergy in Turkish beekeepers: determination of risk factors for systemic reactions. Allergol Immunopathol (Madr.) 34:180–184

Charpin J, Birnbaum J, Haddi E et al (1989) Prevalence of clinical and biological signs of allergy to Hymenoptera stings in general population. J Allergy Clin Immunol 83:229

Cox L, Lavenas-Linnemann D, Lockey RF et al (2010) Speaking the same language: the world allergy organization subcutaneous immunotherapy systemic reaction grading system. J Allergy Clin Immunol 125:569–574

Cox LS, Sanchez-Borges M, Lockey RF (2017) World allergy organization systemic allergic reaction grading system: is a modification needed? J Allergy Clin Immunol Pract 5:58–62

Eich-Wanger C, Muller UR (1998) Bee sting allergy in beekeepers. Clin Exp Allergy 28:1292–1298

Fischer J, Teufel M, Feidt A et al (2013) Tolerated wasp sting challenge improves health-related quality of life in patients allergic to wasp venom. J Allergy Clin Immunol 132:489–490

Freeman TM (2004) Hypersensitivity to hymenoptera stings. N Engl J Med 351:1978–1984

Goldberg A, Confino-Cohen R (2010) Bee venom immunotherapy: how early is it effective? Allergy 65:391–395

Golden DBK (2009) Insect allergy. In: Adkinson NF, Busse WW, Bochner BS et al (eds) Middleton's allergy: principles and practice, 7th edn. Elsevier, Philadelphia, pp 1260–1273

Golden DBK (2015) Allergy to insect stings and bites. World Allergy Organization

Golden DBK, Marsh DG, Kagey-Sobotka A et al (1989) Epidemiology of insect venom sensitivity. JAMA 262:240–244

Golden DBK, Moffitt J, Nicklas RA (2011) Stinging insect hypersensitivity: a practice parameter update 2011. J Allergy Clin Immunol 127(4):854.e1–854.e23

Haeberli G, Bronnimann M, Hunziker T et al (2003) Elevated basal serum tryptase and Hymenoptera venom allergy: relation to severity of sting reactions and to safety and efficacy of venom immunotherapy. Clin Exp Allergy 33:1216–1220

Heilman J. (2010) Urticaria. Wikipedia

Hunt KJ, Valentine MD, Sobotka AK, Lichtenstein LM (1976) Diagnosis of allergy to stinging insects by skin testing with Hymenoptera venoms. Ann Intern Med 85:56–59

Levine MI, Lockey RF (2003) Monograph on insect allergy: American Academy of Allergy, Asthma, & Immunology, 4th edn. Dave Lambert Associates, Pittsburgh

Lockey RF, Turkeltaub PC, Olive CA et al (1989) The Hymenoptera venom study II: Skin test results and safety of venom skin testing. J Allergy Clin Immunol 84(6):967–974

Loveless MH (1962) Immunization in wasp-sting allergy through venom repositories and periodic insect stings. J Immunol 89:204–215

Loveless MH, Fackler WR (1956) Wasp venom allergy and immunity. Ann Allergy 14:347–366

Mauss V (2008) Bionomics and defensive behaviour of bees and diplopterous wasps (Hymenoptera, Apidae, Vespidae) causing venom allergies in Germany. Hautaarzt 59:184–193

Miyachi S, Lessof MH, Kemeny DM, Green LA (1979) Comparison of the atopic background between allergic and non-allergic beekeepers. Int Arch Allergy Appl Immun 58:160–166

Muller UR (2005) Bee venom allergy in beekeepers and their family members. Curr Opin Allergy Clin Immunol 5(4):343–347

Muller U, Spiess J, Roth A (1977) Serological investigations in Hymenoptera sting allergy: IgE and haemagglutination antibodies against bee venom in patients with bee sting allergy, bee keepers and non-allergic blood donors. Clin Allergy 7:147–154

Muller U, Helbling A, Berchtold E (1992) Immunotherapy with honeybee venom and yellow jacket venom is different regarding efficacy and safety. J Allergy Clin Immunol 89:529–535

Pesek RD, Lockey RF (2014) Treatment of Hymenoptera venom allergy: an update. Curr Opin Allergy Clin Immunol 14(4):340–346

Richter AG, Nightingale P, Huissoon AP et al (2011) Risk factors for systemic reactions to bee venom in British beekeepers. Ann Allergy Asthma Immunol 106:159–163

Rueff F, Placzek M, Przybilla B (2006) Mastocytosis and Hymenoptera venom allergy. Curr Opin Allergy Clin Immunol 6:284–288

Rueff F, Chatelain R, Przybilla B (2011) Management of occupational Hymenoptera allergy. Curr Opin Allergy Clin Immunol 11:69–74

Sampson HA, Munoz-Furlong A, Campbell RL et al (2006) Second symposium on the definition and management of anaphylaxis: summary report—second national institute of allergy and infectious disease/food allergy and anaphylaxis network symposium. J Allergy Clin Immunol 117:391–397

Settipane GA, Boyd GK (1970) Prevalence of bee sting allergy in 4,992 boy scouts. Acta Allergol 25:286–291

Severino M, Bonadonna P, Passalacqua G (2009) Large local reactions from insect stings: from epidemiology to management. Curr Opin Allergy Clin Immunol 9(4):334–337

Stoevesandt J, Hain J, Stoize L et al (2014) Angiotensin-converting enzyme inhibitors do not impair the safety of Hymenoptera venom immunotherapy buildup phase. Clin Exp Allergy 44:747–755

Strupler W, Wuthrich B, Schindler Ch (1997) Prevalence of Hymenoptera venom allergy in Switzerland: an epidemiological and serological study based on data from Sapaldia (in German). Allergo Journal 6(Suppl 1):7–11

Author Biographies

Peter A. Ricketti, DO, is a third-year fellow-in-training at the Division of Sleep Medicine, Department of Internal Medicine, University of South Florida Morsani College of Medicine, and James A. Haley Veteran's Affairs Medical Center in Tampa, Florida. He received a B.S. in Biology from Villanova University, Villanova, Pennsylvania, in 2008 (Magna Cum Laude) and Doctor of Osteopathic Medicine from the Philadelphia College of Osteopathic Medicine, Philadelphia, Pennsylvania, in 2012 and completed his residency in Internal Medicine at Rutgers-New Jersey Medical School, Newark, New Jersey, in 2015. In June 2016, he graduated from the Division of Allergy and Immunology, Department of Internal Medicine, University of South Florida Morsani College of Medicine, where he served as chief fellow during his final year. Upon completion of his sleep medicine fellowship in June 2018, Dr. Ricketti plans to begin practice with his father, Anthony Ricketti, MD, in allergy and immunology, pulmonary, and sleep medicine in Trenton, New Jersey. He has authored or co-authored 11 publications, two book chapters, and five abstracts. He is actively involved in clinical research at the Joy McCann Culverhouse Airway Disease Research Center and Allergy, Asthma, and Immunology Clinical Research Unit at the University of South Florida Morsani College of Medicine.

Richard F. Lockey, MD, MS is a Distinguished University Health Professor; Professor of Medicine, Pediatrics and Public Health; Joy McCann Culverhouse Chair of Allergy/Immunology (A/I); and Director, Division of A/I, Department of Internal Medicine (IM), University of South Florida College of Medicine, and James A. Haley Veterans' Hospital, Tampa, Florida.

He received a B.S. degree from Haverford College, Haverford, Pennsylvania; M.D. from Temple University, Philadelphia, Pennsylvania (Alpha Omega Alpha); M.S. from the University of Michigan in Ann Arbor, Michigan, where he trained in A/I and was a Major and Chief of A/I at Carswell Air Force Base, Fort Worth, Texas, 1970–1972.

He has the honor of authoring, co-authoring, or editing over 600 publications and 35 books or monographs with colleagues and has lectured on numerous occasions nationally and internationally. Professional honors include President of the American Academy of Allergy Asthma and Immunology (1992), past Director of the American Board of Allergy and Immunology (1993–1998), and President of the World Allergy Organization (2010–2012). He has served as co-editor or participant of two WHO reports, has served on many journal editorial boards, and has been honored for his research and contributions to medicine and the specialty.

Over 90 physician specialists and 50 international postgraduate PhDs or MDs in basic and clinical research and medicine, many of whom have assumed leadership positions in medicine throughout the world have been trained in the Division.

The Division, its Joy McCann Culverhouse Airway Disease Research Center and Clinical Research Unit core faculty, is staffed by 3 basic, 6 clinical scientists, and approximately 60 other physicians and healthcare professional.

Areas of expertise and research: insect Allergy; allergen immunotherapy; asthma; inflammatory lung diseases; pulmonary fibrosis; comorbid conditions of asthma; and RSV vaccine development.

Printed in the United States
By Bookmasters